OUR BODIES, OUR PLANET

OUR
BODIES,
OUR
PLANET

A Parasite's History of Us

Marcus Hall

REAKTION BOOKS

To my favourite centenarian,
Charles Chauncey Hall, Jr –
in awe and admiration, as always

Published by
REAKTION BOOKS LTD
2–4 Sebastian Street
London EC1V 0HE, UK

www.reaktionbooks.co.uk

First published 2025
Copyright © Marcus Hall 2025

EU GPSR Authorised Representative
Logos Europe, 9 rue Nicolas Poussin, 17000, La Rochelle, France
email: contact@logoseurope.eu

Printed and bound in Great Britain by Bell & Bain, Glasgow

A catalogue record for this book is available from the British Library

ISBN 978 1 83639 107 4

CONTENTS

PREFACE:

A Wormy World

I believe the single most undiagnosed health challenge
in the history of the human race is parasites.

DR ROSS ANDERSON in Valerie Saxion,
Every Body Has Parasites: If You're Alive, You're At Risk! (2003)

My grandmother loved to tell the story of how my own mother, when she was a little girl of about seven years old, was discovered to have parasites. A brief inspection at the toilet had revealed that there were little white worms in her stool. After attending to my mother's infestation and administering the appropriate medicine, everyone agreed that the real culprit was the family dog, an old collie that must have passed on these creepy crawlies to their beloved tomboy, who spent more afternoons exploring nearby vacant lots than she ever did playing with dolls. But upon a visit to the veterinarian, it was pointed out that the dog had no such worms, or at least had recently got rid of them. It was thus assumed that the one at real risk of contagion was the dog.

My mother may have been carrying a few roundworms, such as *Ascaris lumbicoides,* a fairly large intestinal worm that can reach a length of several centimetres, and which apparently did make appearances in suburban Chicago in those days. Or maybe the never-identified worm was just a simple tapeworm (*Taenia* sp.), or segments of it, which is also quite striking and potentially rather long. Yet the most likely candidate for my mother's early experience with her fellow travellers is the pinworm, or *Enterobius vermicularis*, which were common inhabitants of kids' intestines in those days, as they are today. White and stretching a centimetre in length, like the others they are mostly harmless – as well as being highly human-specific. In fact most human intestinal parasites are not readily exchanged with dogs at all, since human and canine parasites are typically quite selective about their hosts – except perhaps for threadworms such as *Strongyloides stercoralis*, but even these are not very visible and not very worrisome. And so the family collie was an unlikely source of my mother's worms. Whether or not my mother was just parasitotropic – had an affinity for parasites – may be less important than simply acknowledging that today, children everywhere are probably not hosting as many parasites as they once did. For better, but possibly for worse, our bodily ecosystems are not as biodiverse as they once were.

✳

MY AIM IS to write parasites back into human history. This book's premise is that over the long run, creatures living in and on us have been altering and even improving our lives through their efforts to improve their own. I travel into the world of our intestines and our scalps to show how widespread these creatures really are, while tracing some of the ways we came to discover this mostly hidden world. Parasites of humans, micro and macro, ecto and endo, come in all forms and sizes and include single-cellular protozoa as well as multi-cellular worms and arthropods, many of which can be readily detected with a little practice. Health records show that today some 40 million Americans feed pinworms, 11 million Germans host roundworms and at least 10 million East Asians carry intestinal flukes. Such creatures range in length from a few millimetres to several centimetres, with one textbook asserting that '*Homo sapiens* rank among the most parasitized of all animals ... our varied habitat and diet and our global distribution exposes us to more infections than any other species.' Little wonder that a well-trained parasitologist can find at least fifty different kinds of parasite on a single person – and many others go unidentified. A more exhaustive count reveals that some 280 different species of intestinal worms alone can live in a typical human gut, and this does not include the many other creatures that burrow in our hearts and livers, between our toes, up our noses and in our armpits. Lice and fleas thrive on your neighbour's children, and on yours, and on you. Ninety-seven per cent of us play host to *Demodex*, a genus of tiny mites belonging to the spider family that snuggle in the hair follicles around our eyes. We are instinctively repulsed by these little freeloaders, but what collateral effects do they have on our lives and lifestyles, culture and arts, politics and dreams?[1]

Creepy crawlies also infest our farm animals and forest creatures, and use both in complicated multiple-host life cycles that may involve us. If you own a cat, there is a good chance that it has already passed on its *Toxoplasma gondii* to you: although such invisible parasites are not supposed to be very dangerous, some researchers have linked them to altering the very ways we think and act. If we classify bacteria and even viruses as parasitic, because many live and depend on their hosts, including us, our parasitic co-travellers number in the hundreds of millions of species, with raw numbers of *individual* parasites numbering in the billions, at least. The latest findings of the Human Microbiome Project are confirming the stunning range and diversity of creatures that call humanity home. In fact, it is now almost a cliché to point out that healthy adult bodies host ten times more bacterial cells than human cells – which is to say that if there are 60 to 90 trillion human cells in every one of us, then there are ten times this number of *non*-human cells being toted around by each of us, all day, every day, at work and play, at home and in bed. *Homo sapiens* is hardly a single species going at it alone, but a veritable superorganism carrying a Noah's Ark of creatures through life's journey.

Born from zoological and veterinary interests, the science of parasitology has gone on to show its relevance to human health, agricultural production, wildlife management and almost every other field linked to the life sciences. In our fundamental activities of breathing and sleeping, eating and digesting, reproducing and ageing, our bodily creatures are interacting with us and with each other. Such interactions involve varying degrees of competition and cooperation – for hurting or helping us and the other creatures that depend on them. Within and between species, individual parasites compete with their hosts for resources, but they also cooperate with them to share and multiply resources. Instead of viewing our intimate creatures as simple pathogens that produce disease, many parasites can be re-envisioned as cooperators and symbionts. Humans and their army of parasites march together down life's path, often for the benefit of each other. If a *pathogen* causes disease, a *parasite* denotes a relationship – and even when parasites become pathogenic, most of them are not as problematic for their hosts as we may assume.

Biologist and social philosopher Peter Kropotkin (1842–1921) was an early proponent of cooperation who believed that evolutionary forces typically motivated individuals to help rather than hurt one another, especially when confronting challenging environments. A century later, microbiologist Lynn Margulis (1938–2011) was still celebrating cooperation as a key biological driver, manifested by such processes as symbiogenesis, whereby a simple cell or sequence of nucleic acids becomes incorporated into a larger, more complicated cell or organism. Margulis and her followers believe that key cell organelles such as mitochondria and chloroplasts originated as free-living organisms that migrated into larger cells, initially acting as parasitic creatures to eventually begin providing benefits to their hosts. In the long run, the proverbial parasite has not often acted as a simple freeloader that weakens or sickens the creatures that host it, but instead has shown itself to be a productive, even cooperative and beneficial organism for diversifying and assisting life on the planet.

It turns out that the first parasites, so named, were actually a guild (or profession) of humans who lived in ancient Greece and were in the habit of depending on wealthy patrons for their food and shelter

– which suggests an unbalanced relationship of taking and not giving. But such parasites were actually compensating their hosts for their meals and services in various ways, such as by providing them with good conversation or entertainment, and were so skilful at doing so that their relationship with their host could continue indefinitely. Successful (human) parasites and hosts were giving to one another in stable partnerships, with each understanding their duties that made their pacts work. Only if parasites lost their ability to entertain or soothe (or their host ran out of resources or lost interest) would their relationship break down, leaving the parasite and host to go their separate ways. Over the centuries, *parà•sítos* – 'eat beside' – was a term extended to the plant and animal worlds, and would eventually return to the human world as a biological metaphor, while losing much of its assumed reciprocity. Humans would be viewed as 'parasites' if they took more than they gave. To be 'parasitic' today means to be a selfish freeloader, but the term's etymology requires us to wonder whether a parasite may often be a vital and productive member in a give-and-take partnership. It is the cooperative nature of parasitism that has such enormous implications for life on this planet, including human life.

Within our human world, one might extend the metaphor of parasitism to the scenario of an irresponsible and often-absent employee, for instance, who is paid the same salary as a more diligent employee who never misses a day of work. Or a 'parasite' might be a university student who sees nothing wrong with slacking off on her studies while lazily relying on the financial support of her parents. Or perhaps human parasitical behaviour is reflected in the infant child who is suckled by his mother to survive and grow. All these parasitic individuals may be consuming more than they are producing. But there are functional and ethical reasons why, in each of these cases, our society may tolerate and even assist those who give according to their abilities and take according their needs. Noted historian William McNeill also leaned heavily on parasitic metaphors in his social interpretations of history, calling attention to upper-class 'macroparasites' who exploited the lower classes, and to colonists who 'parasitized' newly settled lands for amassing untold wealth; McNeill offered explanations for why this kind of parasitism has long existed and is not always simply self-serving. Parasitism fulfills biological and cultural roles, and there may even be such a thing as good parasites or selfish hosts.

With such broad interpretations of parasitism, and multifarious assumptions about the roles that parasites can play, we can appreciate why these creatures merit our attention, and not just because of their tendency to infest other organisms. Parasites play crucial roles in the biological world and also in our cultural world, from food webs and ecosystems to immigration and politics, even to the Anthropocene and Gaia theory. I begin Chapter One by exploring the curious links between parasitism and human health, with long life as a proxy for

'Weren't they great hosts? / Weren't they terrific parasites?', cartoon by Dan Piraro, *Bizarro Comics*, 2011.

sustained good health. If we can suppose that parasites are not always bad for us, then the absence of them may actually pose health risks and even prevent us from living longer. One parasite that recurs throughout this book is the one responsible for producing malaria, a single-celled protozoa called *Plasmodium* that measures a fraction of the width of a fingernail and resides in our bloodstreams and livers, causing much human suffering and death. Yet because some strains of malaria parasite have co-existed with *Homo sapiens* for so long – essentially ever since our species has walked the planet – we must wonder about the other effects it is having on us besides causing illness. In the grand expanse of the last 200,000 years, this parasite has not managed to eliminate us, nor have we or our immunological systems managed to eliminate it. One can then ponder the mystery of how many people today – millions of people – continue to host *Plasmodium* in their bodies but never feel any ill effects. The question thus arises of what these parasites are doing to us if they are not making us ill. How are our lives and lifestyles, even our relationships and landscapes, being modified by this exemplary parasite that is so ubiquitous? The island of Sardinia serves as a starting place to ask these questions, since malaria was once so prevalent there but is today almost entirely gone. Now that the disease has been eradicated from that Mediterranean island, with the last case of endemic malaria being eliminated more than 75 years ago, what has changed, or will change, in the lives of Sardinians, and have such changes all been for the better?

My questions are largely historical, but I also rely on scientific insights to help make sense of any conclusions that can be drawn. Chapter Two tracks the accumulating discoveries of scientists over the years to show that many of the questions they posed a century ago are still relevant today. Harvard naturalist William Wheeler explained in 1928 that 'parasitism is an extremely protean phenomenon, one which escapes through the meshes of any net of scholastic definitions.'[2] Not even professional parasitologists now working in modern laboratories always agree on the definitions of their main creature of study, since our understanding of a parasite can be so elusive. Some parasitologists assume that parasites are simply creatures that take more than they receive and so damage their host. Others counter that many parasites are maligned and misunderstood, and if they cause damage or disease, such activities are merely temporary distractions in a parasite's march towards becoming a cooperative and healthy partner with its host. The decorated molecular biologist and Nobel laureate Joshua Lederberg felt that even our most irksome parasites and pathogens are evolving to become less dangerous and less virulent over time. A disease's natural progression holds enormous implications for ourselves and for the creatures that depend on us.

It may therefore be a parasite's occasional benevolent effects that are its most surprising trait. Tracking the multiple benefits of being parasitized is the goal of Chapter Three, both at the individual and species scale. There is now significant agreement, for example, that a rich load of intestinal parasites can actually help us defend ourselves from certain diseases, or even mitigate such problems as auto-immune disorders thought to stem from our excessive use of antibiotics and parasite scrubs. The *Demodex* mite that lives on nearly all of our eyebrows is no longer considered a harmful parasite that should be eliminated, but a possible commensalist that goes about its activities without causing any damage – or even a mutualist that benefits us by cleansing our sebaceous follicles of harmful microbes. We would probably be in worse shape without our eyebrow inhabitants. Still other parasites may benefit their hosts through more unexpected and indirect means, as by boosting disease resistance, conferring psychological advantages or even promoting human equality, as we will see. Yellow fever, and the viruses responsible for it, have always been a scourge on those who contracted them, but even greater harm typically awaited those who initially avoided being exposed to this disease to then contract it later in life. Those who caught mild yellow fever in their youth built immunities that they could utilize later, as when confronting adversaries and enemies who lacked such immuno-experience. Here, the timing of being parasitized could offer terrific benefits to the host. Indeed, the accepted thesis about the dangers of transporting lethal microbes across oceans in a 'Columbian Exchange', as elaborated by Alfred Crosby and others, gains new understanding when interpreted through the benefits of being parasitized. Old World parasites that unleashed suffering and death in the New World were, from a different perspective, simply compensating their hosts by providing them with crucial immunological advantages. Instead of viewing the transport of microbes as a simple process of spreading disease, the global exchange of microbes can be recast as cooperative projects between parasites and hosts. Hosts have been reaping more benefits from their parasites than they may suppose.[3]

Chapter Four focuses on what happens when people set out to eliminate their parasites, illustrating the often grave consequences to land and life that follow. By the mid-nineteenth century, drinking tonic water was a common malaria preventative, since the antimalarial quinine might be dissolved in liquids to provide a bartender's menu of prophylactic beverages. Yet not only did the *Plasmodium* parasite build resistance to this active ingredient of cinchona bark, but patients might suffer debilitating side-effects from this remedy, made more dangerous when concentrations were increased. In the tenacious struggle to separate a parasite from its host, many medical practitioners were slow to realize that their battery of antiparasitics often produced more harm than cure, especially by eliminating the benefits these parasites were providing. When the battle against malaria turned harder into killing the parasite transmitter, the mosquito, then human health as well as ecosystem health was further jeopardized when organic and inorganic pesticides settled across the countryside. Much of the silence of Rachel Carson's spring stemmed from humanity's battle with its parasites, even though some of these parasites were rather innocuous creatures that produced few or no symptoms. Asymptomatic parasitemia has been the rule rather than the exception in the human experience with parasites.

We can reap additional insights about our parasites by seeing them as a common household creature. Like our dogs and cats, our domesticated lice, fleas and *Demodex* might protect us against dangers and discomforts when we provide them with habitat and food. Maybe we can even feel a sort of empathy towards our whipworms if they help us to avoid Crohn's disease. Traditional Chinese shamans assumed that a few worms in the gut and lice in the scalp were a sign of robust health, especially after observing that infirm and elderly individuals don't carry many of these parasites, or else lose them as their health conditions worsen, with the little creatures typically departing a failing body to seek a healthier one. Thinking like a parasite is the topic of Chapter Five, even if this is a challenging task, realizing that our own minds may never be able to enter those of a protozoa – such as *Toxoplasma gondii*, now found in more than a fifth of cat owners. Such intra-species thinking may none-theless disclose that from the perspective of our fellow travellers, we hosts are the main parasite in this relationship, since the many and varied benefits that we reap from our parasites might mean that they are the ones that merit compensation. And even if it is difficult to pinpoint all the advantages that our parasites provide us, one can still appreciate the complicated interactions flowing between parasite and host. Since most parasites also host their own parasites, a parasite's interactions with its host are more complicated than expected, especially when we real-ize that a parasite's parasites typically host still smaller parasites – with all such parasites aiming to satisfy their hosts, else tensions mount, their unions disband and mutual symbioses disappear. The goal of this chapter is to convince us to show more empathy for our parasites.

One can also ratchet up, rather than down, to explore parasite–host unions at macro-levels of landscape, continent and biosphere. Predator–prey relationships can be envisioned as one species parasit-izing another, with a group of foxes parasitizing while also hosting a group of hares: fox predation improves the cunning and vitality of hare species, even though an individual hare suffers. Individual detriment can provide collective benefit for humans as well. Feedback between an organism and its ecosystemic network reveals crucial dependencies between animate and inanimate worlds at the planetary scale, from

Ionic temple block showing a Greek ruler, probably Arbinas of Xanthos, flanked by helpers, with servants at left and a parasite at right, 4th century BCE.

lithosphere to hydrosphere and atmosphere. We arrive at Chapter Six ready to explore parasitical relationships in a Gaia system and see how *Homo sapiens* interacts with its macrohost, the whole Earth. All creatures, people included, construct niches for making life more viable on this planet. Can our insights about parasitism be harnessed to allow us to understand the human transformation of the Earth into an Anthropocene, the latest geological epoch in which human beings are the primary sculptor? If humans have become Gaia's principal parasite, our species maintains a special interest in keeping Earth inhabitable. When projecting from microbe to macrobe, one must remember that parasites evolve as the host evolves; that individuals may be sacrificed for the benefit of the species; that maladaptive parasites are replaced by more cooperative ones; and that the quest for better human health can usher in better earthly health. If humanity is not providing tangible benefits to the biosphere, Gaia mastermind James Lovelock could be correct in asserting that the living Earth will someday act to remove its most troublesome parasite: 'my fellow humans must learn to live in partnership with the Earth,' Lovelock warned, 'otherwise the rest of creation will, as part of Gaia, unconsciously move the Earth to a new state in which humans may no longer be welcome. The virus COVID-19 may well have been one negative feedback. Gaia will try harder next time with something even nastier.'[4]

We finish our parasitic journey by exploring how we can begin to bring back some of our key parasites. With modernity stripping us of a spectrum of bodily creatures that our ancestors once carried, the act of consuming probiotics or receiving faecal transplants becomes a way of restoring and rewilding some of our former microbiomal diversity, the focus of Chapter Seven. Just as we conserve endangered species in reserves and parks, our bodies can become refuges for rare species useful for preserving our own and our world's health. There may be real merit in reintroducing roundworms or flatworms to our intestines for counteracting the rising incidence of multiple sclerosis, say, or rheumatoid arthritis, Crohn's disease and Type 1 diabetes. Various parasitic worms, assert a team of immunologists, if 'applied in a controllable clinical setting, could relieve inflammatory disease yet minimize the adverse effects of the parasite'. One also wonders if there have been successes in restoring parasites to other creatures, at larger scales. How far has *Homo parasitus* benefited the planet, and how can our human epoch be kept inhabitable? A clearer understanding of parasitism makes us wonder how well we are nurturing our own parasites, and how we can better nurture our planetary host.[5]

My call, then, is not only to better acknowledge and appreciate our bodily organisms, but, more importantly, to appreciate and save our partnerships. We live with other beings, small and large, within and on and around us, including other humans, who depend on us, and us on them. The parasitical record suggests that our co-creatures often bring benefits to us, sometimes as they also bring us grief, individually or collectively. By extension, our human species can bring benefits to Earth, or at least has brought such benefits in the past. We can realize that the human body is a union of mutualists, just as on the planetary scale, the biosphere is a union of mutualists. We can realize that the human body is a union of mutualists, just as on the planetary scale, the biosphere is a union of mutualists. It is from our microbes that we can better learn to live with our macrobes.

1

Sardinians Do It Longer

Sardinia is 'the place with the highest concentration
of male centenarians in the world.'
New York Times, 2012

S ardinia is the stuff of spaghetti Westerns, and in fact, a few of
them were actually filmed there. But if you look carefully at those
silver screens of dusty cowboys and wild horses, rugged hillsides
and lonely railroads, you might just spy a chestnut tree or wandering
goat that hints at their real location in the heart of the Mediterranean.
As this sea's second largest island, which only by chance belongs to Italy,
Sardinia has nurtured mostly poor shepherds and humble farmers,
except for such notables as Grazia Deledda, 1926 Nobel laureate in literature, or Antonio Gramsci, founder of the Italian Communist Party.
Long an agricultural people who avoided the coasts, Sardinians in the
past few decades have witnessed modernity arrive rudely and abruptly
to their azure beaches, with summer tourists from Rome or Frankfurt
seeking pristine sunshine and luxury hotels. Only in winter can today's
visitor still glimpse the old life, and sit down at a bench in the village
square next to a leather-skinned gentleman who with some coaxing may
tell you how it once was – if you can understand him – for he probably
won't be speaking Italian, but rather one of the island dialects, languages
so ancient and distinct that many of their words pre-date Latin. Much
more than other peoples, Sardinians and their habits are indigenes: of
the place, born here, even evolved here.

While conversing with the tidy old man on the bench, we can try to envision the world in which he grew up. In his youth, tourists came rarely to his island, as it was long a net exporter, not an importer of people searching for a better life. Back then, only the most intrepid of visitors considered coming to Sardinia's shores to explore its deeper mysteries. One such visitor was D. H. Lawrence, whose *Sea and Sardinia* recounts a memorable 1921 train journey that found the villagers to be almost as seasoned as the local *casu marzu* and *porceddu* – tangy sheep's milk cheese and roast pork. Lawrence's passages offer clues to that earlier land and life: 'Yes, the steep valley sides become almost gorges, and there are trees,' noted Lawrence, but the visitor does not see thick forests, as imagined, 'but scattered, grey, smallish oaks, and some lithe chestnuts. Chestnuts with their long whips, and oaks with their stubby boughs, scattered on steep hillsides where rocks crop out.' Staring out the window, Lawrence describes his

> train perilously winding round, half way up. Then suddenly bolting over a bridge and into a completely unexpected station. What is more, men crowd in . . . Some have the sheep-skin tunic, and all wear the long stocking-cap. And how they smell! of sheep-wool and of men and goat . . . Full of coarse life they are: but so coarse! The handsome fellow has his sleeved waistcoat open, and his shirt-breast has come unbuttoned. Not looking, it seems as if he wears a black undervest. Then suddenly, one sees it is his own hair. He is quite black inside his shirt, like a black goat.[1]

Today, a journey across Sardinia still reveals a wild land brimming with extremes: it is the Mediterranean's most inaccessible major island, and its oldest geologically. High numbers of rare or unusual plants and animals find refuge here, making it a biodiversity hotspot. Its people are Europe's most genetically distinct ethnic group – harbouring rarer sequences of DNA than any other Europeans, excepting perhaps northern Scandinavia's Sami people – as there has been little immigration to the island since early colonizers arrived on its shores some 8,000 years ago. This limited mixing of genes means that Sardinian people display distinct physical traits, such as wide shoulders and broad noses – as well as unusual health conditions. Although many islanders enjoy such hardy advantages as dark, sun-resistant skin and remarkable endurance, they also suffer inordinately from rare genetic illnesses such as coeliac disease, Wilson's disease, thalassemia and favism, all of which demand restricted diets to alleviate the associated indigestion, nausea and headaches that come with these conditions. 'It is a strange, strange landscape,' Lawrence surmised, 'as if here the world left off.' No wonder anthropologists now flock to the island. Human geneticists have made it a favourite study site. Demographers increasingly descend on its shores.[2]

Demographers come to study population statistics, for another Sardinian anomaly is human longevity. Sardinians are some of the longest-lived people on the planet. Whether their genes or their

Long-lived Sardinians: four of nine siblings of the Melis family
whose combined age in 2012 was 818 years.

lifestyle, or both, these islanders have apparently discovered the secret to living longer. A 2005 study concluded that in Sardinia, some 208 people out of 100,000 typically live past the age of one hundred, a rate of longevity more than twice that of other Western countries. The mountainous interior of Sardinia is even more spectacular in promoting long life, nurturing twice the number of centenarians than the island as a whole. Only scattered villages in faraway Costa Rica or Japan's Okinawa archipelago vie with the interior of Sardinia for stacking up human years. In fact, in 2012 one Sardinian family of nine brothers and sisters were awarded a Guinness World Record for their combined age of 818 years. Such impressive statistics are trumpeted in popular reports and tourist brochures, which note that Sardinia's Methuselah hotspots can count one centenarian for every two hundred people. This figure is roughly twenty times higher than in England and thirty times higher than in the United States, where a recent census shows that just one person in 5,786 lives beyond their hundredth birthday. Not surprisingly, the rush is on to drink from Sardinia's fountain of youth.[3]

Blue Zones

National Geographic writer Dan Buettner has written a series of bestsellers about the world's 'Blue Zones', which he describes as regions on a world map that can be outlined in blue to demarcate places of longliving people. Buettner's main goal has been to answer the pressing question about why Blue Zone people live so long: how do they do it? What do they eat, and how much exercise do they get? How hard do they work? And how can the rest of us learn from them for living longer ourselves? Not surprisingly, Sardinia plays a central role in Buettner's books.

Genes or environment or both: Buettner initiates his investigation by asking experts on ageing and geriatrics to specify longevity's main factors, which range from healthy food to regular exercise, tobacco abstinence, mental stimulation and favourable genes – especially as shown by having long-lived blood relatives. Yet as Buettner continues to search for clues to longevity, the one issue about Sardinians and their island that surprises him perhaps more than any other is that, before the 1950s,

> Sardinia had looked more like a backwater than a centen-
> arian's Shangri-La. Poor hygiene, poor water quality, and a
> scarcity of water led to rampant infectious diseases. Dysentery,
> plague, tuberculosis, malaria, and diarrhea killed many young
> Sardinians.[4]

Tellingly, Buettner's deeper investigations corroborate Lawrence's 1921 picture of a once disease-riddled island. Lawrence had observed, for example, that the windows at many of Sardinia's railway stations were covered with insect screens as a way to keep out mosquitoes and the diseases they carried: 'The shallow upland valleys, moorland with their intense summer sun and the riverless, boggy behaviour of the water breed the pest inevitably.' One begins to understand that dis-eased and difficult conditions were once the norm, not the exception, for Sardinians and their children.[5]

And so by the logic of chronology, one realizes that some of these children of yesteryear who were surrounded by filth and disease were exactly the ones who would grow up to become today's centenarians and longevity record holders. Buettner calls attention to this puzzle of unsanitary surroundings fostering so many vigorous individuals, a point also noted by William Henry Smyth, another travel writer of Sardinia who, a century before Lawrence, wrote that 'it is surprising that with such inconvenient residences, and uncleanly habits, the natives should remain so generally healthy as they do.'[6]

To Buettner's good fortune, several local centenarians agreed to meet up with him, together with other village elders who would volun-teer insights about their lives and lifestyles, diets and genes. He discov-ered from Giuseppe Mura, age 102, that bread is one of his favourite island staples, but not fish, for example. Buettner found that his other long-lived informants consumed only moderate amounts of meat, poured in plenty of mastic oil, and often chased it all with deep-red wine – while surrounding themselves with light-hearted company and plenty of conversation. The best recipe for a longer life, Buettner would conclude, is to 'Eat a lean, plant-based diet accented with meat; Drink goat's milk; Put family first; Celebrate elders; Take a walk; Drink a glass or two of red wine daily; Laugh with friends.' An interesting omission here, however, is that in his pointers for long life, Buettner does not recommend that we emulate Sardinian centenarians' childhoods of poor hygiene surrounded by infectious diseases.[7]

Sardinia's Disease History

Although one recent eight-authored study in *Experimental Gerontology* focusing on Sardinia's longevity puzzle could only conclude that that there may be dozens of factors that can explain the islanders' long lives, it seems that the most interesting observation before us is the simple fact that today's oldest Sardinians were once surrounded by threatening health issues. Indeed, by looking deeper into this island's past, one discovers that another crucial Sardinian superlative is precisely its dramatic experience with human disease, both infectious and hereditary, with an inordinate exposure to one disease in particular: the disease of 'bad air' – or *mal-aria*, its Italian namesake. Although the cruel disease of malaria still rages in many parts of the world, it deserves special mention when recounting the history of Sardinia. Even more than the plague or dysentery or tuberculosis or typhus, Sardinia's most notorious health concern has always been malaria, a disease that has formed much of the backdrop to the day-to-day challenges of the Sardinian way of life.[8]

Sometimes crowned the queen of diseases, malaria produced (and produces) recurring fevers, headaches and nausea in all ages, with these symptoms being most severe in the earliest years. Generations of Sardinian children have been debilitated and devastated after being bitten by malaria-carrying mosquitoes. Disease experts long considered Sardinia to be one of Europe's most malarial places as tabulated by mortality and morbidity, as well as one of the temperate world's last malaria strongholds. As Maurice Le Lannou, one of the most insightful anthropologists to study Sardinia, explained in 1941, 'malaria in Sardinia holds a special place amongst all other diseases: it affects not a limited group of unfortunates, but an entire population.' Rates of malaria infection here were comparable to or surpassed those of the tropics, and in many respects this disease was even more dreadful in Sardinia because its temperate climate produced seasonal, rather than continual, exposure to mosquitoes, thereby preventing islanders from building up the degree of immunity to the disease that was enjoyed by some of their tropical counterparts. It took tightly sealed housing and powerful insecticides before malaria could be eradicated from the island, finally being definitively extinguished by 1950. By contrast, most other Europeans saw their malaria problems disappear a generation or two earlier than did Sardinians. Likewise, the rest of the temperate world witnessed most of its malaria vanish by the 1920s and '30s, as in the United States, where malaria finally disappeared from the southern states and California.[9]

In those days, malariologists – malaria scientists – often demarcated malaria outbreaks on maps with dark shades, or Black Zones, as they might be called. Sardinia was typically portrayed in the darkest of tones. During the first half of the twentieth century, the Rockefeller Foundation's International Health Division viewed the island as one of its crucial worldwide test sites for finding better ways to combat malaria, which meant that a battery of newly fabricated antimalarial drugs and mosquito-killing techniques saw their first application there. The widely used arsenic–copper insecticide Paris Green was sprayed across the island in the 1920s to kill mosquito larvae during this chemical's first European trials in controlling malaria. And following the invention and introduction of the more powerful DDT during the Second World War, the Rockefeller Foundation directed an unprecedented,

Sardinian man guiding oxen, 1948.

all-out DDT-spray campaign across the island from 1946 to 1950, with the goal being to completely rid Sardinia of malaria-carrying mosquitoes. It is fair to say that before our own day's wave of foreign health advocates and research demographers began arriving on the island, Sardinians of two generations ago witnessed their own cosmopolitan parade of malariologists and mosquito eradicators coming to test the day's latest remedies on these shores.[10]

Sardinians still continue to celebrate their island's post-war 'liberation' from malaria, as in 2010 when a convention was organized in the main city of Cagliari to commemorate the sixtieth anniversary of 'the war won against the scourge that for centuries determined the island's history'. One could say that a hundred years ago, it was in fact malaria that made this island famous – not its picturesque beaches and strong cheese, and much less its spectacular longevity. In tracking the island's experience with disease, there can be no denying that today's longest-living Sardinians grew up as children under harshly malarial conditions.[11]

A Spurious Correlation?

And so, like a lightning bolt in dark sky, the key question flashing before us is to wonder whether there can be a strange or unusual relationship between two of Sardinia's most famous traits: severe malaria and remarkable longevity. Is it possible that Sardinia's twentieth-century experience with malaria can be linked to its twenty-first-century production of centenarians? Or, is this occurrence of disease and longevity mere spurious correlation instead of logical causation? High disease incidence would seem to promote *shortened* longevity. Yet by twisting common sense here, one can wonder whether yesterday's high numbers of malaria sufferers and today's high numbers of very old shepherds is more than coincidence. We need to take up the issue of how being exposed to a dreaded disease and then surviving it might be related to longer life. If we consider humanity's protracted, brutal experience with malaria – which continues to punish enormous swathes of humanity today – are there any conceivable reasons why an endemic disease should provide collateral benefits to the people who experience it, be these physiological or cultural, even extending their lives disproportionately?

If one focuses closely on the malady that is malaria, one realizes that it is caused by a group of tiny parasites that live part of their lives in the human liver and bloodstream, where they seek to thrive and multiply before being passed on via a mosquito to another person's blood to thrive and multiply in their new host. One important point to realize is that malaria is not always deadly, now or in the past. In fact, only about one out of five hundred people who catch malaria today end up dying from it. Even in centuries past, most malarial carriers

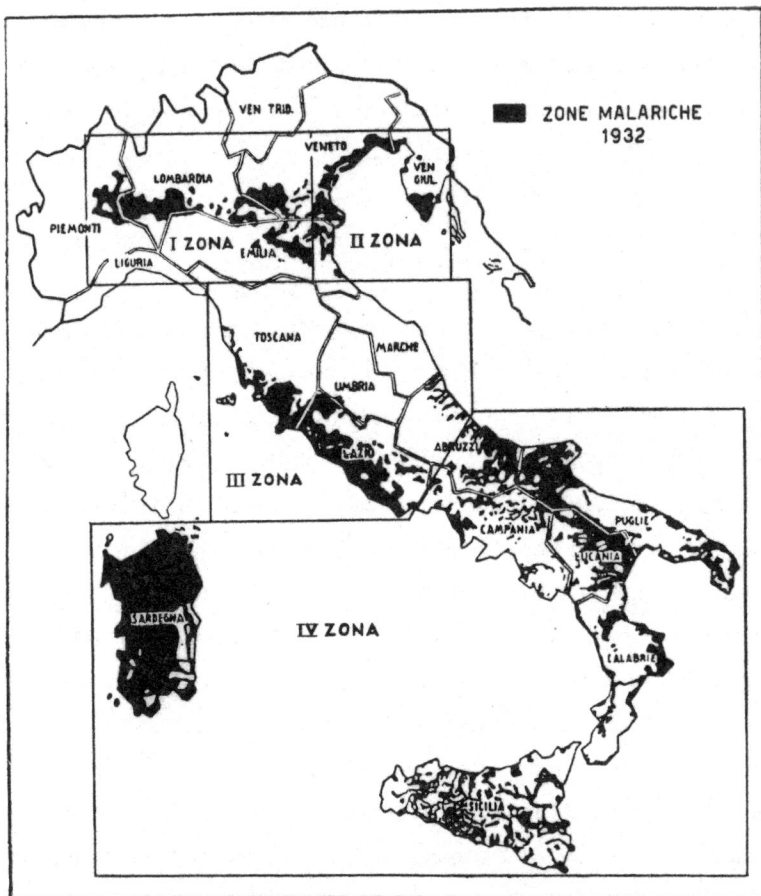

Map of Italy showing malaria zones in black, 1932.
Sardinia is the island coloured mostly black.

without access to modern medical remedies did not succumb to malaria. It may be that the parasites responsible for malaria, like all parasitic creatures, maintain a crucial interest in keeping their hosts alive in order to keep themselves alive. In reviewing the biology of parasites, one finds that through complex and sometimes convoluted ways, via spatial, temporal and behavioural processes, evolutionarily successful parasites work to ensure that their hosts will also be successful, else the parasites will not survive. A 'successful' creature means one that manages to continue living and reproducing for long periods over many generations. From the perspective of malaria's single-celled parasite called *Plasmodium*, it would be very happy if its host could go on living and serving as a reservoir for rearing, multiplying and transmitting more plasmodia.[12]

For just a moment, then, we might look past the terrible suffering and mortality caused by malaria to consider if there can be any short- or long-term advantages to the host if they contract a key parasitic disease. At the very least, we must ask why hosts have not been able to rid themselves of their age-old parasites. Why hasn't *Homo sapiens* been more successful at expelling its malaria parasites over all of the millennia it has hosted them? In the evolutionary long run, it would seem that a host should be able to shake off a debilitating parasite, else the host will suffer inordinately and possibly disappear – or perhaps at some stage the two can reach a tolerable equilibrium. One can therefore understand that a crucial question is to wonder whether parasites are, in fact, giving something back to their host so as to be sure that they continue receiving the host's benefits. To rephrase our query: might parasites be part of the reason why malaria's former Black Zones are now longevity's Blue Zones?

Black Zones within Blue Zones

Of course, it is almost unspeakable to suggest that malaria can bring anything but suffering to the human species. This disease consistently ranks as a leading killer that, along with AIDS and tuberculosis, is among humanity's most deadly infectious diseases, currently slaughtering some 400,000 individuals each year, many of them children. To put that number in perspective, the dreaded coronavirus that accelerated across the world in the 2019–23 pandemic would go on to kill more than ten times that number. However, malaria has been sowing its own fields of death every year, year after year, for millennia. Just two decades ago, malaria's annual death toll reached 1 million individuals annually. Considering the long sweep of human history across the past 200,000 years of our species, some tabulators believe that malaria has killed more individual human beings than any other single disease. Multiplying malaria's annual proportion of victims by the number of years *Homo sapiens* has walked the Earth produces estimates that number in the billions.[13]

Yet such dramatic numbers provide exactly the rationale for exploring the strange correlation between malaria and longevity. In fact, when searching across the globe and identifying the diseases that once surrounded today's other longest-living persons, one can find still more Black Zones overlapping Blue Zones. In both Okinawa and Costa Rica, where there are also high concentrations of long-lived people, medical historians reveal that malaria was once rampant, likewise receding in the early twentieth century and largely disappearing by the 1950s and '60s. Records show that, as in Sardinia, high percentages of today's oldest Okinawans contracted malaria as children. The story is repeated in Costa Rica. Archives there reveal that a century ago, Costa Ricans were surrounded by severe malaria, particularly in the western coastal lowlands where most of today's centenarians grew up. We can then ask again whether Sardinia's, Okinawa's and Costa Rica's former high rates of malaria and current high rates of longevity are simply strange and unexplainable correlations. Perhaps.[14]

Possible causal relationships between malaria and longevity are certainly complicated and even convoluted, since there are so many ways that one can begin to link long life with early exposure to this disease. To speculate loosely for a moment, contracting malaria and then surviving it may have conferred higher antibody resistance to other diseases confronted later in life, for example. From a different perspective, perhaps the project of draining swamps and channelling rivers, which was so widely implemented for shrinking mosquito habitat and battling malaria, also served to protect humans from other dangerous mosquito-borne diseases such as yellow and dengue fevers or encephalitis. Perhaps former malaria remedies, which involved a spectrum of medications and convalescent therapies, also provided some protection against several other childhood illnesses. Or maybe the heightened economic growth that followed malaria eradication in each of Sardinia, Okinawa and Costa Rica some sixty to seventy years ago provided collateral health benefits or even psychological advantages for those growing up in less developed economies. Longevity researcher Valter Longo and colleagues have found evidence that low-calorie diets and periodic fasting, as was common during pre-war, malarial Sardinia, may inhibit disease-forming processes later in life. It may therefore be the combination of physiological and cultural factors that explains the paradox of finding malaria's Black Zones overlapping with longevity's Blue Zones. More brainstorming would allow one to formulate a long list of conceivable medical, psychological, economic and political reasons why the exposure to malaria may in some places, in some instances, contribute to longer-term human health.[15]

One certainty is that malaria has to date proven to be one of humanity's most complicated infectious diseases, and continues to elude medical scientists more than a century after the discovery that female *Anopheles* mosquitoes can transmit between humans five species of single-celled protozoan *Plasmodium*. Because of continually changing antigen signals given off by these parasites, infected humans have difficulty building up definitive immunity to them after exposure, but nonetheless may acquire resistance (or partial immunity) to them if infected periodically. In highly malarial areas where people are repeatedly infected by hungry malaria-carrying mosquitoes, many of these individuals can become fully asymptomatic carriers of the parasite, and enjoy near full immunity. Although public health experts currently estimate that some 200–300 million people contract malaria and then survive it each year, even greater numbers of people simply carry this parasite and show no adverse effects or clinical symptoms. That is to say, despite our twenty-first century's massive and ongoing malaria control campaigns, which involve pharmaceuticals, pesticides, biocontrols, bed nets, land management, CRISPR technologies and sterilized mosquitoes, together with a host of other high- and low-tech strategies, numerous species of *Plasmodium* continue to live and prosper within much of humanity – with many of them failing to exert much harm on their hosts.[16]

To consider numbers again, around 3 per cent of people on Earth will contract malaria this year, with thankfully a much smaller proportion of those dying from the disease. Yet, as noted, even larger proportions of people will carry *Plasmodium* but show few or no symptoms. Exact proportions of asymptomatic, parasitaemic malarial individuals

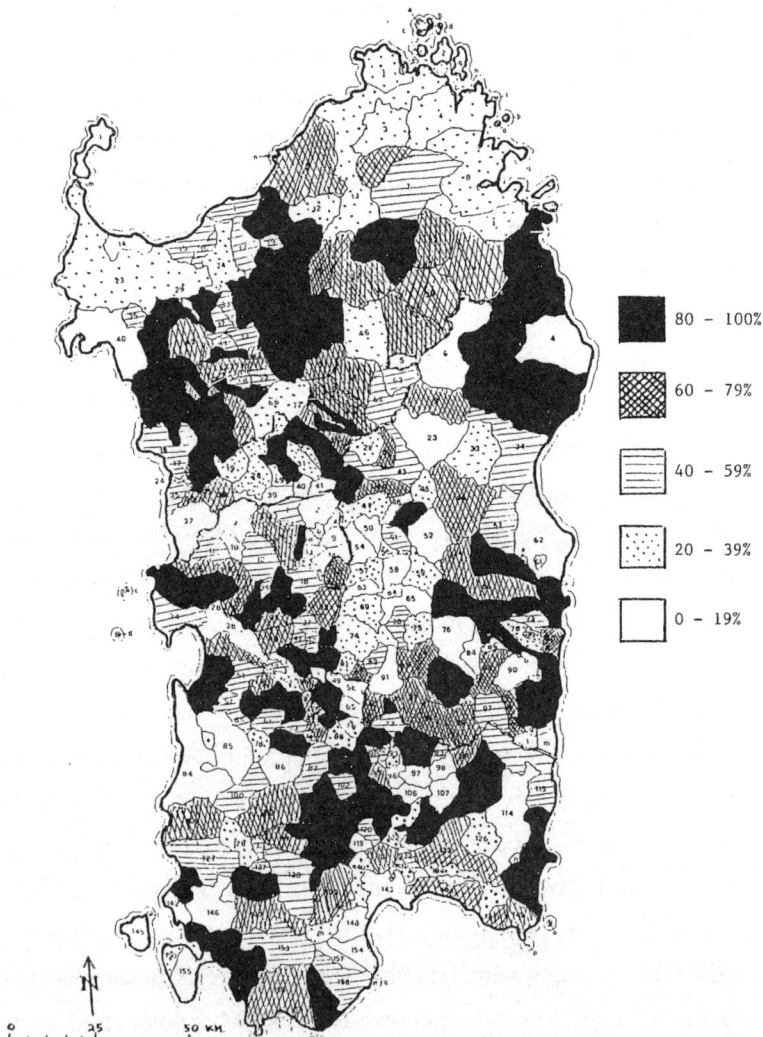

Map of Sardinia showing the prevalence of
malaria by region, *c.* 1935.

are hard to come by, and may vary considerably from region to region. For example, malaria researchers at sites in the western Amazon found that some 30 per cent of the local inhabitants were asymptomatic *Plasmodium* carriers. Another malaria study focusing on villages in Zambia concludes that 'over half the population may be parasitaemic and show no obvious clinical symptoms.' Researchers in Nigeria report that some three-quarters of a large sample of women tested positive for the malaria parasite even though none of them exhibited malaria symptoms. Such studies suggest that very large numbers of people currently harbour *Plasmodium* parasites in their bodies, particularly in the tropics. To then extrapolate these numbers to the global human population who live in the tropics would indicate that upwards of a half-billion people today are carrying plasmodia in their bodies. Yet the majority of those people feel almost no ill effects from this parasite. And so, if these parasites are not actively causing disease, the main mystery before us is what all of these parasites are up to.[17]

If we turn the clock back a hundred years to the days when malaria was even more widespread across the planet, we can confirm that many of yesteryear's parasite carriers were the same people who went on to lead normal and sometimes very long lives. Even though a very few individuals experience long-range debilitating side-effects from malaria, doctors report that most malaria carriers exhibit no serious long-term consequences. We must therefore ask again whether *Plasmodium* parasites may not be as dangerous as we assume, and in fact may be giving their human hosts an extra push in life's race.[18]

Malaria in Human Affairs

Beyond malaria's potential links with human longevity, we can be more certain about this disease's other effects on human society. Over the ages, *Plasmodium*'s association with humans has shaped our genes and our bodies, our behaviours and our lifestyles, even our architecture and landscapes, acting as a surprisingly powerful historical agent.

From the impersonal perspective of evolutionary principles, malaria acts fiercely at the first stages in life, meaning that over generations, its effects on the human species have been particularly dramatic by acting hardest before a person can reach adulthood to nurture the next generation. In other words, malaria's childhood susceptibility means that its victims do poorly in passing on their genetic heritage. When biological or cultural defences arise against the disease – however minuscule – such defences are quickly multiplied and favoured in the next generations. One geneticist deems malaria to be 'the strongest known force for evolutionary selection in the recent history of the human genome'. Another scientist rejoins that malaria is the 'strongest evolutionary selective force in recent human history'. A group of public health experts describe how, for example, the simple practice in the early 1900s of placing mosquito screens across open windows found rapid adoption across Europe and North America – and across the tropics – after it was observed how useful such screens were in limiting the spread of malaria. One can therefore appreciate that malaria's biological pressures exert cultural consequences, all mediated across generations.[19]

From the long perspective of a thousand human generations, it is fair to say that all of us living today are malaria survivors. Despite malaria's ongoing and tragic death toll, most of humanity today has found a way to co-exist with the parasite. Thanks to our ancestors' genetic adaptations combined with their accumulating cultural ingenuities, most people have developed myriad known and unknown ways of coping with malaria, otherwise we wouldn't be here. The long co-existence of *Homo sapiens* and *Plasmodium* means that both organisms waltzed down their evolutionary paths together, continually reacting to the other's moves. As humanity's immune systems and cultural practices developed better ways to cope with the parasite, whether by producing better antibodies or constructing mosquito-proof housing, the parasite's and mosquito's own defence mechanisms in turn developed better ways to adapt to our remedies – such as by building up resistance to our drugs and pesticides or becoming better at slipping past the screens in our windows. This co-reactionary process has been called an 'evolutionary arms race' whereby *Plasmodium/Anopheles* attacks are mirrored by human counter-attacks, with all such attacks being fine-tuned as the generations progressed.

But might this arms race have occasionally evolved towards a détente? If the parasite launches a softer offence, the host might itself respond with a softer defence, with both species benefiting. The long-term result of launching progressively softer attacks can result in more peaceful co-habitation – between nations, as well as between humans and their parasites. It turns out that co-evolving counter-attacks can indeed lead to less virulence: the fact that *Plasmodium* and humanity are both alive hundreds of millennia after their first encounter suggests that there has been significant progress towards peaceful co-existence.[20]

In Sardinia, such parasite–host co-existence has produced a range of biological and cultural adaptations. Because *Plasmodium* spends part of its time living within and deriving energy from human red blood cells, some of malaria's most direct effects are manifested in the chemical composition of human blood. In fact, at least a dozen human blood anomalies are traceable to the advantages of disrupting some of the functions of red blood cells that *Plasmodium* depends upon. The notorious sickle-cell condition, commonly associated with many African populations and their descendants – and also unusually common in Sardinians and their progeny – is one such genetic condition stemming from the advantages that sickle-shaped blood cells provide in inhibiting *Plasmodium* development. That is to say, if one lives in a malarial environment, it is advantageous to be a heterozygous sickle-cell carrier. Or, as one medical scientist describes it, 'the malarial parasite appears to have exerted positive heterozygote selection for retention of otherwise-deleterious and disease-associated polymorphisms affecting erythroid cells.'[21]

There is a similar explanation for the presence of favism, mentioned earlier, which limits a person's ability to digest fava and other broad beans, since individuals carrying this trait are unable to produce adequate amounts of G6PD, an enzyme that helps break down the alkaloids contained in these beans. Because this enzyme is also useful to *Plasmodium*'s own metabolism as it feeds on the human red blood cell, people lacking the ability to produce G6PD actually have a degree of resistance to *Plasmodium* and may therefore enjoy a healthier life in malarial areas. Even though G6PD-deficient individuals can become violently ill after consuming a plate of fava beans, they are less susceptible to the ill effects of malaria than their non-G6PD-deficient neighbours. Little surprise, then, that Sardinia has one of the world's highest rates of favism – 30 per cent compared to the world average of 8 per cent – and that fava beans (a traditional Mediterranean foodstuff) are currently banned in cultivation projects across the island. After malaria was definitively eradicated from Sardinia three generations ago, only the negative effects of G6PD-deficiency and sickle-cell trait are manifested there today. These examples demonstrate how islanders' blood chemistry as well as their dietary habits and even their crops have all been modified by a strain of parasites that measure just a hundredth the thickness of a fingernail.[22]

Plasmodium Landscapes

Yet genes are just one level on which malaria acts. As shown by the presence of window screens or fava-free diets, human culture in its myriad manifestations is also shaped by exposure to the dramatic debilitator that is the malaria parasite. The attentive traveller to Sardinia today will detect the parasite's cultural imprint at nearly every turn. Town planning, habits and rituals, housing designs, even clothing styles have all been modified by this parasite as people struggled to avoid its periodic fevers and deal with its cruel deaths. A century ago, so ubiquitous and recurrent were malaria fevers – *le febbri* – that most islanders considered them to be inevitable. Like sleeping with buzzing flies or eating mouldy bread, malaria was a nuisance that was worse in some years than others, but was mostly unavoidable. The islanders could not so much prevent the disease as learn to live with it. In their effort to stay clear of the parasite, they built their dwellings far from the boggy marshes or up on breezy hillsides where the 'good air' would blow, often clustering their houses to form hill towns. Shepherds practised reverse transhumance, which found them leading their flocks by winter to lowland pastures when the pestilential airs abated. Dress style was likewise moderated by the disease, as when islanders fended off insalubrious airs (and so the mosquitoes and parasites they carried) by wearing thick garments even on hot summer days so as to minimize the exposure of bare skin. In calling attention to the heavy robes draped over Sardinian peasants, Lawrence noted, 'They say it keeps off the malaria. The men swathe shawls round their heads in the same way.'[23]

Sardinians also had malaria in mind when naming their most sacred shrines, such as Il Santuario di Nostra Signora Bonaria – the Sanctuary of Our Lady Good Air – located in the outskirts of Cagliari. Legend has it that, in the fourteenth century, a Spanish ship was caught in a violent tempest offshore, whereupon the sailors began jettisoning cargo to keep their ship afloat. After one particularly heavy trunk was thrown overboard, the storm suddenly abated, allowing the sailors to avert catastrophe and keep their ship on course. When local friars later found a heavy trunk washed up on the city's beach, they opened it to find a small statue of the Madonna, which they named after the adjacent Colle di Buenaire, or Hill of Good Air. Not only did the Aragon king then order a castle and basilica to be constructed on the hill to commemorate the miraculous event, but Spanish conquistadors voyaging to South America would name Argentina's main city, Buenos Aires, after this Sardinian site. In these times before any germ theory of disease, it was the *good air* that Sardinians searched for and cherished, thankful that it would waft away the enervating drafts of *bad air*. By the turn of the twentieth century, Pope Pius x would proclaim the Madonna Bonaria the patron saint of Sardinia for her role in bringing good

Il Santuario di Nostra Signora Bonaria, Cagliari, Sardinia.

fortune – and healthy air – to the island. In subsequent decades, Paul VI and John Paul II likewise made pilgrimages to Cagliari's Madonna on the hill. From the architectural to the economic, from the physical to the spiritual, Sardinia's pesky *Plasmodium* commanded even popes to come to the island.[24]

The parasite's influence has been stamped across Sardinia's countryside in a thousand other ways, too, beyond influencing village locations, shepherd habits and shrine names. Beginning in the 1910s and '20s, government-sponsored water reclamation projects across the island sought to drain or else fill mosquito-breeding marshes, while straightening stream courses wherever possible. Eucalyptus seedlings were imported from Australia for planting in row after row across lowland areas to desiccate mosquito habitats. Mosquitofish (*Gambusia affinis*) were imported from North America and introduced into streams and ponds to consume mosquito larvae. Today's photographs of Sardinian river valleys taken from the same locations as their old,

Statue of the Madonna of Bonaria, the patron saint of Sardinia,
standing next to her church.

black-and-white counterparts demonstrate the dramatic changes the
landscape has undergone.

The island's all-out mosquito-killing campaign of the late 1940s
also altered its economic landscape. When 32,000 DDT sprayers and
applied entomologists found precious seasonal employment from
1946 to 1950 by exterminating mosquitoes, they injected their precious

Draining Sardinian landscapes near Cabras, 1948.

wages into a cash-poor economy to produce multiplier effects on local prosperity. As one elderly Sardinian later reflected in an interview about that DDT-spraying era, villagers in those days began 'carrying coins in their pockets'. It seems that Sardinia's intensive pesticide campaign, directed by the Rockefeller Foundation and costing millions in United Nations relief funds, served to jump-start a lethargic economy – as much by injecting cash into a largely subsistence way of life as by helping rid the island of its malaria scourge. From the ethnographic to the

DDT sprayers in Sardinia, 1948.

economic, the tentacles of *Plasmodium* parasites reached into the deepest crannies of Sardinian society.[25]

Beyond Pathos

Humanity's intimate relationships with *Plasmodium* parasites therefore reveal why we need to recount not just histories, but co-histories: human stories entwined with those of our fellow travellers. Alongside our domesticated animals and our cultivated plants that foster human civilizations are the creatures who live in or on us, some of them lethal but others innocuous or possibly beneficial. Far from being a simple disease agent, a malaria parasite does not merely infect the human body, treating it as so much habitat and foodstuff for reaping nutrients to grow and multiply before travelling to the next unsuspecting host. This simple exploitative motive might be the strategy of a *Staphylococcus* bacterium or a *Giardia* protozoan or a coronavirus. Instead, the *Plasmodium* parasite grows through developmental stages within humans, its intermediate host, as well as within mosquitoes, its definitive host, to circulate and develop in both organisms as it grows, reproduces and gets itself transmitted to other hosts. The human liver is *Plasmodium*'s main site of dormancy before it moves to blood cells to undergo accelerated growth. The parasite's developmental stages of sporozoites, schizonts, trophozoites and gametocytes are simply labels in a complicated life cycle illustrating how this micro-organism grows in the human body before entering and developing further in the gut of a mosquito, to finally be injected into its next human host and repeat the process. It is fair to say that this parasite's overarching goal is to live with its hosts and adapt to their habits so that it can go on living. Although *Plasmodium* takes from us, it does not take everything, else it would not survive, since we would not survive. This parasite may in the end have little concern for an individual human life, but it is deeply concerned about the fate of the human species.

One can also appreciate the distinction between a pathogen and a parasite. The former makes us sick; the latter lives in or on us and *may* get us sick. 'Pathogens' comes from *pathos*, the pain and suffering that these creatures inflict on their victim by feeding on it and multiplying to go on to their next victim. Parasites, however, result from *relationships*, whereby creatures live in or on their hosts for access to habitat and reaping nutriment and may inflict very little pathos. *Plasmodium* can produce fever, fatigue, nausea, dehydration, headaches, even blood clots and eventually death. But in its many dormant and non-virulent forms, and in less susceptible hosts, malaria's parasite is not pathogenic, not all of the time. Just as *Escherichia coli*, our ubiquitous intestinal bacteria, are usually harmless within our intestines, as by aiding our digestion under normal circumstances, they can under other bodily conditions become dangerously pathogenic by secreting toxins or invading other organs to make us very ill indeed. Like *E. coli*, *Plasmodium* is sometimes but not always pathogenic to its hosts.[26]

The science of parasitology is concerned not only with the medical, veterinary and agricultural questions of parasites and their hosts, but increasingly with ecological and evolutionary questions that explore relationships between these co-travellers and their environments. When describing organismal interactions, ecologists traditionally distinguish such relationships into the categories of mutualism, commensalism and parasitism, which reflect some of the complexity of how two species can interact with one another. A diligent science student will remember that

Malaria
(Plasmodium spp.)

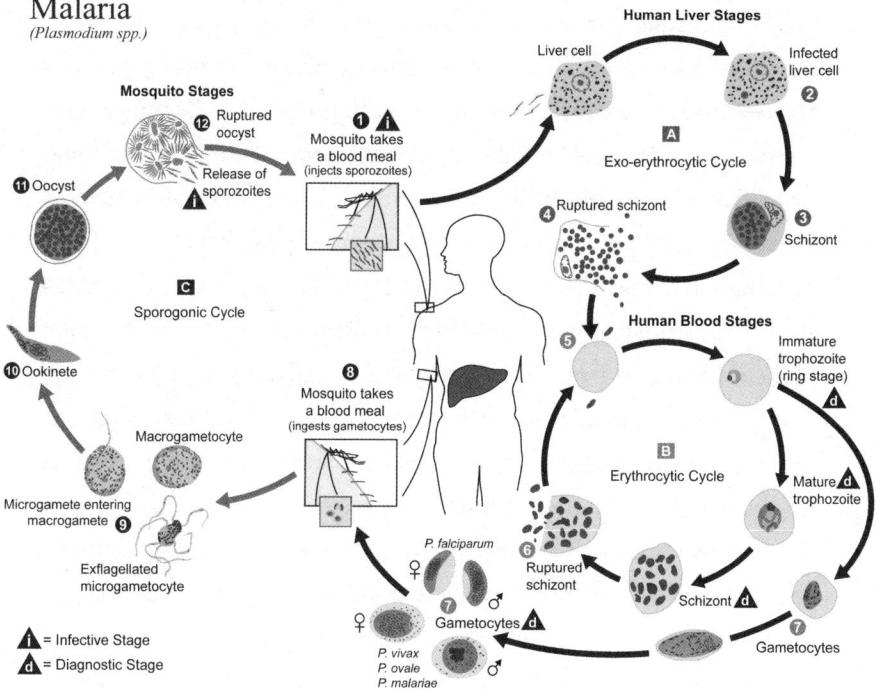

Malaria parasite cycle.

mutualism is a partnership that benefits both creatures, such as the pilot fish that grooms a shark's skin in exchange for protection by the shark. Commensalism is a partnership that benefits one member while leaving the other unaffected, such as sea barnacles that harmlessly attach themselves to a shark's skin in order to reap better feeding conditions. And for its part, parasitism classically denotes a partnership whereby one benefits to the detriment of the other, as exemplified by a shark's intestinal worms found feeding on their host's hard-won nutriments.

But more meticulous observations of these organismal associations generally show that these three categories blend into one another, since there is often no clear indication of which of the partners is reaping the most advantage in their unions. In the case of a shark's intestinal worms, there is now evidence showing that they absorb and concentrate toxic heavy metals within their host, allowing their posterior segments to be expelled out of the digestive tract, thereby serving to detoxify the sharks of dangerous poisons. A relationship that was once considered to be classic parasitism is now viewed as mutualism: each creature assists the other. More and more relationships involving internal organisms are now labelled simply as 'symbiotic', with the precise advantages and disadvantages that each member derives from the relationship still being worked out. Mistletoe, the plant that grows on trees and reaps nutrients by tapping into its host's vascular system, may in a new light be providing benefits to the tree by attracting nesting birds that help disperse the tree's seeds. Mice artificially infected with the protozoan parasite *Trypanosoma* or the roundworm *Trichinella* actually gain useful weight, apparently due to metabolites that they induce their hosts to secrete. Such observations reveal that many parasites are offering their hosts something in return for the benefits they are reaping.[27]

In those instances in which a parasite is clearly feeding selfishly and detrimentally on its host, then their relationship simulates a pathogen injuring its host, or perhaps a predator feeding on its prey. In fact, some parasites are considered to be 'predatory parasites' if they inflict significant damage on their hosts. As pioneering ecologist Charles Elton wrote in 1927, 'it is best to treat parasites as being essentially the same as carnivores [that is, predators], except in their smaller size, which enable them to live on their host . . . the resemblances between the two classes of animals are more important than the differences.' Yet parasites can be distinguished from predators in that the latter typically have a greater choice of creatures that they can attack and consume. In the macro-organismal world of bears, wolves and lions, such predators can seek various prey, from bite-size mice to full-meal rabbits. Many parasites, though, typically prefer to live in or on just one or a few hosts, and reap the available resources, having co-evolved with their host's distinct internal chemistries, temperatures and rhythms.[28]

So-called obligate parasites are those creatures that cannot survive without a host, and many such parasites are very selective of their hosts. Any measure of biodiversity should therefore account for all the creatures living on or in a particular host species, since a single host may be a walking bundle of co-dependent parasites. Consequently, threats of extinction faced by a single host species are also faced by its parasites: one study about co-extinction predicts that should five endangered North American mammal species go extinct, 56 of their parasite species would follow them into oblivion. Biodiversity enthusiasts should therefore take note that the world's often forgotten and ignored parasites represent enormous numbers of species – at least half of all species on Earth, according to conservative estimates. Other researchers expand such numbers, claiming that parasites constitute 70 to 80 per cent of all the world's species. The vast majority of such parasites lurk unknown, unobserved and unlabelled within the bowels of larger species.[29]

Respecting our Co-Travellers

A fuller understanding of parasitism in human affairs therefore requires that we look closely at host–parasite relationships. Viewing a malaria parasite and its human host as engaged in a vicious arms race is too simplistic, since each creature may be giving more to the other than meets the eye. Parasites have increasingly been found to carry out not only helpful but vital functions for their hosts. If parasites act as pathogens or predators here, they serve as commensalists or mutualists there. Not simply a relationship of struggle and competition, parasitism can produce many kinds of collaboration and cooperation. Observations in the laboratory and field are showing that parasites can often be more successful and more fit by pleasing rather than perturbing the hosts upon which they depend.[30]

Before peering deeper into the intimate lives of parasites in the next chapter, we might remember that World Malaria Day is commemorated each 25 April, emphasizing *Plasmodium*'s cruel pathogenicity that inflicts widespread suffering on humanity. Ongoing malaria control measures involving millions of funds and armies of healthcare workers continue to develop vaccines, administer medicines, spray insecticides, identify biocontrols and distribute mosquito bed nets around the world. Untold numbers of people suffer and die each year from this horrific disease. The desperate campaign to identify better anti-malarial drugs, propagate sterile mosquitoes and disrupt *Plasmodium* transmission attest to the pain and anguish that can be traced to this little protozoan. Ours is a dire quest to eradicate malaria.

Yet against this backdrop, there is case after case of species aiding one another, even cases of seemingly antagonistic organisms helping each other in their struggle against shared enemies and common harsh elements. There is the parasitic intestinal worm absorbing dangerous toxins from its shark host. There is the parasitic mistletoe luring birds to disperse the seeds of its arboreal host. And there is the amazing observation that many of today's longest-living humans suffered from malaria as children. In the quest to unravel the mystery of Sardinia's remarkable longevity, we need to learn more about the role of parasites in human history. In the following chapter, we discover that parasitology has long been a contested science, being much more than the study of unequal relationships whereby greedy parasites do little more than exploit hapless hosts.

2

Adventures in Parasitology

The history of modern medicine, so far as infectious diseases
are concerned, is nothing more nor less than the history of
parasitology in its broad sense.

ASA CRAWFORD CHANDLER,
Animal Parasites and Human Disease (1918)

A big parasite deserves a big name. *Umingmakstrongylus palli-kuukensis* is an arm-length worm that lives its adult life coiled inside the lungs of a musk ox, the wild ox that grazes the arctic prairies. Eric Hoberg knows how this parasite got its name, for he is the one who named it: *Oomingmak* ('the bearded one') is Inuit for musk ox; *strongylus* relates to a genus of nematodes, or roundworms (Greek: *strongulos*, round); and Paulatuk is a small village in Canada's Northwest Territories, near where he first collected it. As a research parasitologist, Hoberg recounted to me how he had to travel 80 kilometres by snowmobile and then, with a special government permit, locate and shoot a large musk ox before butchering it to extract the beast's lungs, finally removing a few lungworms from that vascular mass. It was 'a big bloody mess,' he said.

Right there in the snow and freezing cold, I reached in under the ribs to cut out the lungs before pulling them out and packing them into a beer cooler for the ride out. At those temperatures, the cooler was to keep the lungs from freezing. Back at our shop, it took quite some time to locate and remove the worms, and then begin untangling two nice specimens, male and female, to finally get them separated.

A big, calm man with a full beard who looks like he really could field dress a musk ox at thirty below zero, Hoberg explained that lungworms are found in all sorts of mammals, from caribou to bison to Dall sheep and snowshoe hares, as well as domestic dogs and cats. Most animals infected with lungworms, he added, appear to be quite healthy.

Hoberg's arctic journey was intended to reveal the degree of taxonomic overlap between species of lungworms and the species of mammals who host them. Because species' evolutionary lines can be diagrammatically linked with each other through branching trees, Hoberg wanted to see how closely such trees of lungworms shadowed those of their hosts. His published evidence showed that indeed, as mammals evolved away from their primordial ancestors, so did their lungworms evolve from theirs.[1]

Hoberg carries out most of his computational work near Washington, DC, at the laboratory of the U.S. National Parasite Collection, a venerable facility that has been serving parasitologists since 1892. As director of this national repository, Hoberg is in charge of some 100,000 'lots' of parasites, or batches of organisms collected in separate samplings. Many of these lots hold so many critters that no one has ever bothered to count them. The millions of embalmed creatures to be found here in carefully stacked glass slides, tubes, vials and mason jars, many filled with glycerin or ethanol, are the fruits of four generations of collectors, with the first specimens courtesy of scientists working out of the early Smithsonian Institution along with the then Bureau of Animal Industry. Worms, or 'helminths' in parasitological parlance – flat, round, long, short, hook, pin, tape – mostly measured in millimetres and visible to the naked eye, are the main strength of the U.S. National Parasite Collection. Yet this *potpourri* of human and animal co-travellers also includes microscopic protozoa such as *Giardia* mounted on glass slides, together with more visible fleas, chiggers and ticks trapped in vials, as well as hand-sized flukes and other parasitic miscellanea bobbing in glass jars, all captured, pinned out and preserved in various developmental stages.

Alongside those token musk ox lungworms, a few of the other pride-and-joy specimens of this collection include multiple-metre tapeworms (*Taenia saginata*), once thriving in human intestines and now coiled inside bottles; a mounted pair of canine kidneys, one normal and the other atrophic, which had been hollowed out by kidney worms (*Dioctophyme renale*). There are vials upon vials of white roundworms (*Parascaris equorum*) extracted from foals and still considered an enemy of horse breeders. And poised near the main door is 'Fritz', the whole cat head bottled in solution that, while bringing giggles to visiting school groups, showcases severe mange caused by subcutaneous mites (*Notoedres cati*). This cabinet of horrors, housed and protected in the U.S. Department of Agriculture's unassuming Building 1180 at Beltsville, Maryland, is one of the world's premier parasite collections – a veritable Noah's Ark of wriggling creatures now frozen in preservatives. The

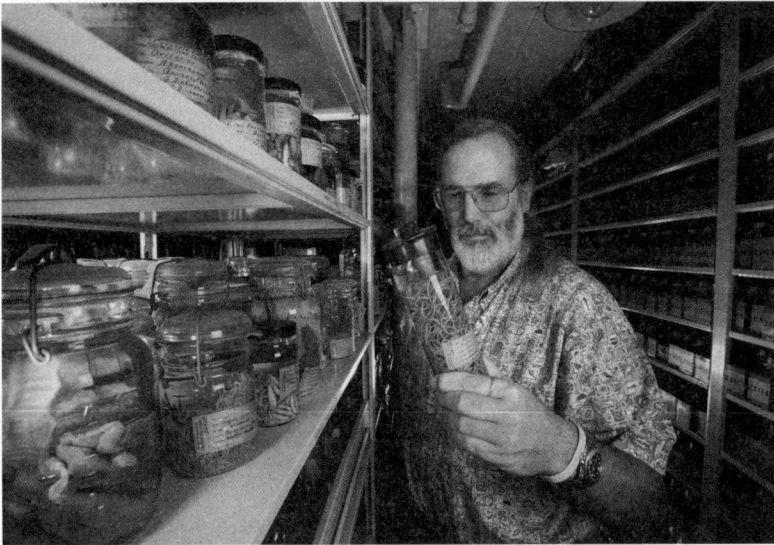

Eric Hoberg, chief curator of the U.S. National Parasite Collection, examining roundworm specimens from raccoons.

millions of prokaryotes and eukaryotes, roundworms and flatworms, annelids and arthropods, all keyworded and catalogued in shelves of jars and vials were collected in the days when men were men and parasites were parasites.

Parasitology in the late nineteenth century was becoming a bona fide science, as medical and agricultural researchers around the world found ever more organisms living in and on people and their domestic animals, at the same time as they sought better remedies to expel or eradicate such creatures. For the USDA and for universities in Nebraska, Illinois, Ontario and Quebec, together with key medical and veterinary institutes in Europe, Russia, South America and Asia, parasites were of interest largely because they were linked to disease and suffering, assumed to rob farmers and livestock grazers of productivity and generally weigh down the spinning wheels of progress. Or, as Henry Ward of the Nebraska Zoological Laboratory explained in 1895, 'the average number of parasites in domestic animals means a considerable draft on the energy of the host, a tax which, were the parasites absent, would be expressed in so much marketable flesh.'[2] Research on the theoretical questions of parasites involving ecological and evolutionary principles of the type asked by Hoberg had to wait decades. But judging by articles in biological journals and trends in scientific societies, most parasitological research began falling out of fashion by the 1950s and '60s, when it was supplanted by work in genetics, molecular biology and biochemistry. The parasite collection in Beltsville, once well funded and supporting dozens of researchers – many of whom would rise to serve as presidents of the American Society for Parasitologists – today maintains a full-time staff of two. Although parasites are recently making a renaissance as a subject of study within various biological subfields, parasitology is distinctly lagging in today's annals of science. One wonders what brought on the stunning rise and curious fall of parasitology.

This chapter turns to the insights of early parasitologists to see what they can reveal about how parasites modify and have been modified by human history. Identifying crucial moments in parasitological science gives us insight into how humans and parasites have been changing each other's lives, and how such changes have not always been for the worse.[3]

Of Stools and Scientists

In the spring of 1980 the Rockefeller and Josiah Macy Jr Foundations sponsored a major scientific conference in New Orleans on 'The Current Status and Future of Parasitology'. Attending were leading virologists, bacteriologists, protozoologists and immunologists of the day, some of whom still clung to the title of 'parasitologist'. In his keynote address, Rockefeller Foundation Health Sciences director Kenneth Warren declared that

> It is doubtful that anyone present, no matter of what persuasion, believes that the field of parasitology is in a healthy state, enjoying the well-being of more than adequate funding, attracting the finest students, and functioning at the forefront of modern scientific methodology and ideas.

His remarks, which certainly brought murmurs and stirs in the audience, were not taken lightly by conference delegates, although many of them could not deny that parasitology had fallen from its earlier florid state. Raising their hands, several individuals mentioned that the specialization of parasitology into its many subdisciplines was making the mother science obsolete: as one participant remarked, the many kinds of parasite, from microscopic to macroscopic, and their many modes of transmission, meant that 'you couldn't put them all together in a common framework if you wanted to.' Warren also recounted his own challenges a few years earlier in sponsoring a joint meeting of diverse experts to discuss parasitic diseases of plants, of animals and of humans: 'Can you imagine a group of plant pathologists talking to human disease specialists and immunologists?' Their questions were simply too different,

Warren implied, their techniques too varied, their methods of analysis too disparate. Still, many in that New Orleans audience agreed with Warren that holistic parasitological study nonetheless continued to play a vital role in advancing biological and medical knowledge.

Chiming in was one Eli Chernin, a Harvard microbiologist who recounted the shortcomings of a recent summer seminar he convened at Woods Hole titled 'The Biology of Parasitism'. He explained that the seminar was designed to teach a range of parasitological techniques, from basic identification to cloning and chromosomal staining, all led by renowned practitioners. Yet Chernin complained that he 'came away feeling it is just as important to know how to do a proper stool examination as it is to know how to produce a monoclonal antibody'. Could parasitology's rising focus on faecal assays be diverting too much attention away from more rigorous science? Or as Warren countered in his own comment,

> What bothers me ... is that the emphasis on the importance of stool examinations and identification of parasites has had a negative influence with respect to attracting people to the field. We need both stool examinations and hybridomas. Biological research is a continuum, and all aspects of it are important.[4]

Perhaps most bothersome to the conference-goers was that there seemed to be little agreement on the core questions of their science. The respected Paul C. Beaver, one of the leaders of the field, emphasized the theoretical challenges facing the science. 'I have been unable to find an acceptable brief definition of parasitology,' Beaver declared that day, 'nor have I been able to construct one. Parasitology is the scientific study of parasites, but this leaves us with the task of defining the terms *parasite* and *parasitism*.' He pointed out that although it seemed appropriate to classify many kinds of bacteria and fungi as parasites, such organisms were not generally included as subjects of parasitology. Beaver nonetheless counselled that parasitology should be considered an amalgamation of studies about parasitic protozoa, worms, insects and ticks, with the last two groups being important vectors (that is, transmitters), or roughly the combination of protozoology, helminthology, entomology and acarology. Beaver was making it clear that parasitology was much more than attention to wriggling white worms, especially of the varieties found inside gastrointestinal tracts: a parasite could be an organism that reaps benefits from another creature, an organism that inflicts harm, or an organism that simply lives in or on another organism. Following the early days that dwelled on identification and laboratory preparation, parasitology had expanded to encompass questions on a range of organisms, their methods of propagation and their effects on hosts and on the larger world. After a lifetime of peering through microscopes and sorting through stool specimens, Beaver could tell a parasite when he saw one, but he couldn't craft a good definition of one.[5]

Even in centuries past, parasites were contested creatures. Writing in Diderot's *Encyclopédie* in the 1770s, Louis de Jaucourt and Louis Daubenton described parasites as those 'who creep into good homes to find a well-spread table'. They added that in ancient Greece, a parasite was sometimes despised, but also viewed in positive light, such that all men saw themselves possessing 'a deity to whom they were indebted for the productions of the earth'. Moreover, 'almost all the gods had their parasites.' In referring to parasites of plants, the *Encyclopédie* sounds more modern, describing these creatures as 'species of weeds that grow on trees [because] they live and feed at the expense of others'. The Encyclopédists showed by their definitions that parasites were more than just creatures looking for free handouts, and by their entries revealed that such beings could be human as well as non-human. A century earlier, when Florence's Francesco Redi wrote his influential *Osservazioni . . . intorno agli animali viventi che si trovano negli animali viventi* (Observations of Living Animals that Are Found Inside Living Animals), these bodily creatures were just that – animals found living inside other animals – with no mention of 'parasites' or 'parasitism'.

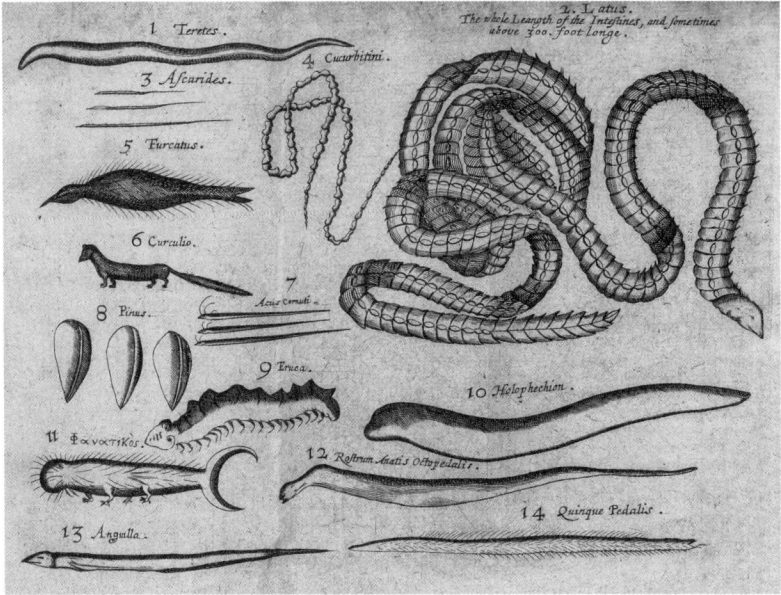

A variety of internal worms of humans as depicted in William Ramesey,
Helminthologia (1668).

Linnaeus in his classification schemes would single out worms – *Vermes*
– as one of six broad categories of organism, but he never referred to
any of them as parasites. As late as 1859, Louis-Benoît Picard's famous
comedy *The Parasite; or, The Art to Make One's Fortune* – translated
from the German, *Der Parasit oder Die Kunst, sein Glück zu machen*
(1803), itself translated from the French, *Médiocre et rampant* (1797) –
never mentioned anything but a human protagonist.

Only later would our internal organisms be considered 'parasitic'
because they simulated humans who were parasites. A literary scholar
who has traced the changing meanings of 'parasite' reveals that not until
well into the nineteenth century was this term generally applied to
non-human creatures, since the term's prototype referred exclusively
to certain humans and their behaviour, rather than the much later
view of humans acting like parasitic creatures: it is 'not the sponging
human, but the biological parasite' that is the more recent phenomenon, explains Anders Gullestad. In short, only 150 years ago did our
understanding of parasites morph from the human to non-human,

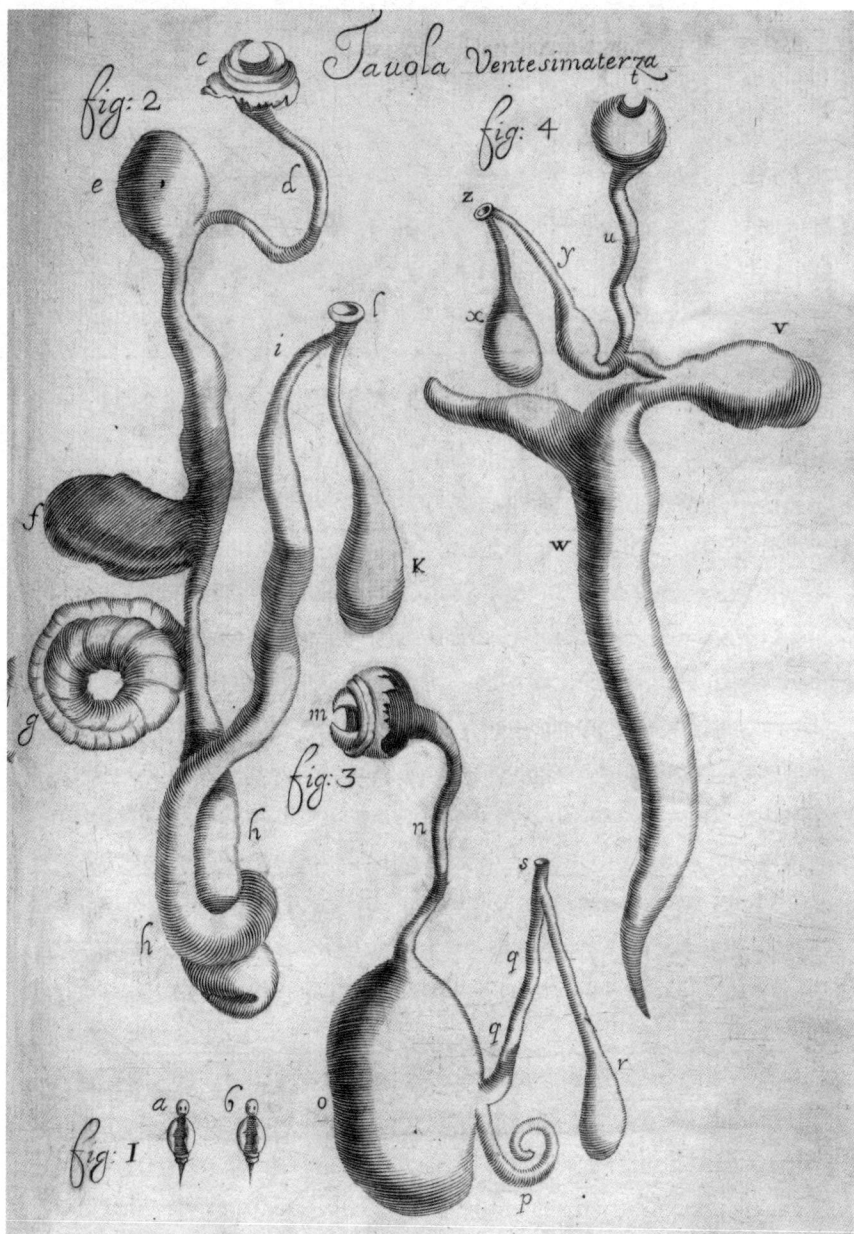

The inner organs of an octopus, from Francesco Redi,
Osservazioni . . . intorno agli animali viventi che si trovano negli animali viventi (1684).

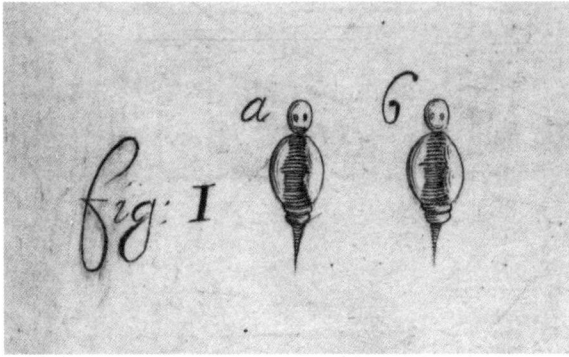

Detail of 'living organisms' found inside an octopus.

and only in our own recent decades did we deem certain humans to
be *metaphorical* parasites.[6]

Despite the growing tendency to view our internal creatures as
freeloaders and, after Victorian naturalists such as Ray Lankester, as
evolutionary degenerates – or creatures that had 'devolved' from higher
forms – did some scientists seek to view parasites on their own terms.
By 1875, when Belgian zoologist Pierre Joseph van Beneden published
his classic *Les Commensaux et les parasites dans le règne animal* (translated as *Animal Parasites and Messmates*), non-human parasites were
becoming bona fide subjects of zoological study. Yet even van Beneden
could not resist anthropomorphizing this class of creatures when he
taught that 'the parasite installs himself either temporarily or definitively
in the house of his neighbor; either with his consent or by force, he
demands from him his living, and very often his lodging.' Here, a parasite was characterized by its propensity to freeload, rather than by its
preferred habitats, for example, or methods of reproduction, means
of transmission or effects on hosts. Man had made parasites in his
own image, and then set out to study these creatures for what they
could tell him about the natural world. No wonder there were differences of opinions about which characteristics of parasites made them
parasitic.[7]

By the late nineteenth century, parasitology had become an established scientific field, following on the heels of the science of

helminthology – or the study of internal worms – or what was sometimes called 'entozoology' – the study of internal animals. More scientists were at this point emphasizing that worms were just one kind of creature living in or on another creature, and indeed such co-inhabitants could be very small, to the point of disappearing under the magnifying glass. More importantly, many such parasites were being linked to human and animal diseases or agricultural blights; that is, parasites were no longer considered *symptoms* but *causes* of illness. Internal creatures were no longer seen to arise through spontaneous generation. With the development of better microscopes and Louis Pasteur's and Robert Koch's germ theory of disease, investigators began searching for still more parasites while setting out to track their life cycles between hosts, which sometimes involved intermediate carriers. If a parasitological disease's aetiological agent could be identified, subsequent investigations usually aimed at finding ways to kill or eradicate it. Nobel prizes were won or lost over deciding who first elucidated a parasite's accurate life history. There is still a simmering debate, for example, over who definitively identified and explained that malaria stemmed from a parasite transmitted by the mosquito, with Italian scholars much more willing than others to give credit to both Giovanni Battista Grassi and Ronald Ross, rather than the Englishman alone.[8]

The historian David Grove, who produced a thick tome about early investigations in human helminthology, calls attention to a growing division in the late nineteenth century between parasitologists of animals and those of humans. Grove notes that medical texts at this time were employing the term 'medical zoology', reflecting how health practitioners were using the knowledge of zoologists to understand and improve human well-being. The USDA's early parasite collection continued this division of animal from human, with the Bureau of Animal Industry's mandate focusing on veterinary and agricultural interests, even though many of its leading investigators would eventually make the medical jump to the nearby Hygienic Laboratory, a precursor to the National Institutes of Health. It seems that there was more money and better facilities over at the medical institute, even though many of the research critters remained the same.[9]

But no sooner was parasitology an established science than its investigators began finding the need to defend their field from critics who considered it too descriptive or too ill-defined. In his presidential address to the American Society of Parasitologists in 1932, Maurice C. Hall countered that 'the discovery of a new star is no more important, scientifically, than the discovery of a new tapeworm . . . Parasitology needs no apology.' Apparently, many investigators were finding Hall's field to be too nebulous to warrant a respectable place in the echelons of the scientific temple. Yet Hall's audience that day undoubtedly stood by him when he asserted that 'the science of coprology or scatology is, at least potentially, as much of a science as mathematics, astronomy, physics [or] chemistry.'[10]

Thickening Parasitological Plots

Maurice Hall may be considered representative of a cadre of well-trained, hard-working parasitologists in the early twentieth century who were using their insights for expanding farm production and improving animal and human health. Hall had risen in the Bureau of Animal Industry to become chief of its Zoological Division, in charge of its enlarging parasite collection that was transferred in 1930 from downtown Washington, DC, to its spacious quarters in nearby Beltsville. The primary mission of the division's brand of pragmatic parasitology was, in so many words, to kill parasites. Hall's search for better anthelmintics – deworming medicines – consumed years of administering various noxious substances to sheep, cows, horses and dogs in solid, liquid and gaseous forms. The main challenge of developing a better worm remedy was finding a substance that could kill (or harm) a worm but did not immediately kill (or harm) its host. Determining an anthelmintic dose's optimal frequency, potency and quantity typically required repeated trial-and-error measurements on dozens of test animals. Hall's eulogizers recounted that 'no method of administration of anthelmintics was dismissed by him as worthless until it had been subjected to critical tests.'[11]

As an aside, it might not be surprising to learn that in his animal experiment work, Hall had scrapes with anti-vivisectionists. In one of his dog deworming trials, for example, he found that the rather toxic oil of Chenopodium failed to remove their parasites, whether injected intravenously or intramuscularly. In defending his work against a newly restrictive law proposed by the Vivisection Confrontation League, Hall justified his research under oath by stating that 'I am as fond of dogs as any proponent of this bill, [and] I have done far more for dogs than any of them have.' In fact, Hall had found Chenopodium oil very useful for other deworming needs, such as for ridding hogs of roundworms. Interestingly, certain groups of Native Americans had for centuries apparently extracted this oil from American wormseed (*Chenopodium ambrosioides*) for purging themselves of intestinal worms.[12]

One of Hall's most wide-reaching discoveries turned out to be carbon tetrachloride, which could be fed to domestic animals for cleansing them of troublesome nematodes, including hookworms. In his trials, Hall followed in the footsteps of his predecessor, Charles Stiles, another noted parasitologist and the first custodian of the Department of Agriculture's parasite collection that one day would be curated by Eric Hoberg, who had observed impressive deworming results in farm animals with such marginally palatable substances as thymol, copper sulphate and even creosote and gasoline. Amazingly, carbon tetrachloride, which is today classified as a potent liver toxin and Group B2 agent ('probable human carcinogen'), didn't seem to harm grazing animals, but it did a wonderful job at scouring their insides of slithering inhabitants – so much so that after Hall's discovery, American farmers began pouring gallons of the stuff down the throats of their livestock.[13]

It required little imagination but much risk to consider human medicinal applications of carbon tetrachloride. So convinced was

Hall of the wonders of this novel anthelmintic that he chose his own body as the site of its first human trials, initially infesting himself with North American hookworm (*Necator americanus*), and then tracking its population in his own faecal samples after swallowing sequential 3-cubic-centimetre doses of the liquid. Thereafter, the Rockefeller Foundation took over where Hall left off, fine-tuning carbon tetrachloride doses on chimpanzee subjects. In all likelihood, assorted snake oils and worm elixirs in those years also contained significant quantities of this compound, perhaps even after its toxic properties became more apparent. But it is certainly an indictment of carbon tetrachloride's even more problematic precursor treatments when one reads in a 1923 report that this substance 'has been given to over 100,000 patients and has so far proved more effective, cheaper and less unpleasant in its effects on the patient than any of the drugs heretofore in use against hookworms'. To satisfy the curious, records show that Maurice Hall himself lived to the mature but not ripe old age of 57.[14]

With regard to the ongoing hookworm campaign in humans, certain well-respected parasitologists in the interwar years became disillusioned with the strategy of pursuing complete eradication of these parasites, especially following the massive 1910–14 hookworm campaign organized by the Rockefeller Foundation in the southern United States. Although the Foundation would declare this campaign a success, hookworms would continue to be found persisting in southern soils and in people's intestines for many years, attracting still greater efforts to eradicate them. By 1942 a few critical voices, such as Justin Andrews, Director of the Georgia Department of Public Health, became more adamant in opposing the strategy of eradication. Andrews pointed out that hookworm infections were often light and subclinical, maintaining that the goal of hookworm control should be to eliminate the disease rather than the parasite. His view was that hookworms were mostly harmless inhabitants of human bodies – creatures that in many cases did not cause disease and were probably not eradicable anyway – meaning that the best strategy was to divert eradication efforts into more pressing health issues: hookworms were unavoidable and simply not worth the fight. Supporting his view is today's worldwide estimate of hookworm carriers, currently numbering between 500 to 700 million, but with just 3,000 of them annually succumbing to any parasitological effects. Hookworms are apparently not as deadly as the Rockefeller Foundation and its proponents had assumed.[15]

One begins to realize that the parasitological community has long been split over how much harm is inflicted by parasites, and whether most of our writhing co-inhabitants should simply be tolerated instead of expelled – since so much collateral damage in trying to expel them might be avoided. Innumerable parasite remedies have shown themselves to be more harmful than the parasites. Thymol, the primary anti-hookworm remedy in the days before carbon tetrachloride, killed more than one patient who did not follow the strict protocol required for its administration. In 1914 a leading veterinary parasitologist, B. F. Kaupp, defined a parasite simply as 'an organism, animal or vegetable, that lives upon another organism, animal or vegetable', and made no mention of a parasite's potential harmful effects. Perhaps Kaupp's veterinary perspective allowed him to locate parasites in most of his own patients – most of whom were quite healthy – so that these wriggling creatures seemed to him to be mostly normal and benign co-travellers. But other investigators maintained that the only good parasite was a dead parasite. Maurice Hall penned a whole farmer's manual patterned on military tactics: his *Control of Animal Parasites* (1936) portrayed doctors as soldiers, drugs as weapons and health remedies as offensive manoeuvres: 'the fight with and the destruction of hostile forces, whether human, verminous, insect, or bacterial, is war.' Clearly, war was on the minds of everyone in those days, so it is hardly surprising that warfare's desperate tactics were finding their way deep into the parasitology labs.[16]

To be sure, Kaupp's radical suggestion that parasites might be harmless co-inhabitants had already been asserted by earlier experts, notably by the nineteenth-century Belgian parasitologist Pierre Joseph van

THE CAMPAIGN AGAINST THE PINWORM

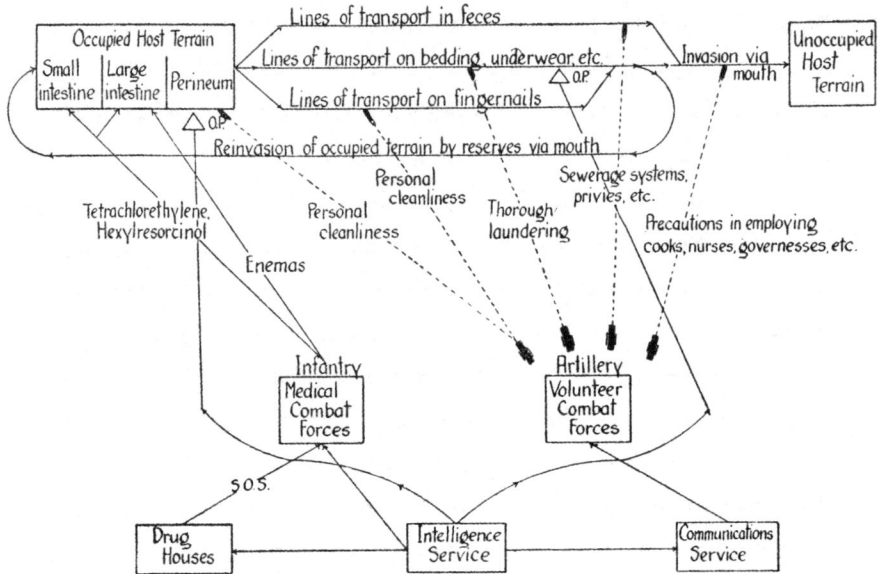

Lines of transport in feces

Occupied Host Terrain

Lines of transport on bedding, underwear, etc.

Invasion via mouth

Unoccupied Host Terrain

Small intestine | Large intestine | Perineum

△ aP.

Lines of transport on fingernails

△ aP.

Reinvasion of occupied terrain by reserves via mouth

Personal cleanliness

Sewerage systems, privies, etc.

Tetrachlorethylene, Hexylresorcinol

Personal cleanliness

Thorough laundering

Precautions in employing cooks, nurses, governesses, etc.

Enemas

Infantry
Medical Combat Forces

Artillery
Volunteer Combat Forces

S O S.

Drug Houses

Intelligence Service

Communications Service

'The Campaign Against the Pinworm', from Maurice C. Hall,
Control of Animal Parasites (1936).

Beneden. His painstaking observations of these micro-creatures in the laboratory and field, with a special focus on marine creatures, convinced him to advance not one but three categories of organism living on hosts, categories that persist in biology textbooks to this day. It was van Beneden who proposed that mutualists benefit themselves as well as their host, that commensalists (or 'messmates' as a British translator offered) benefit themselves while having no negative effect on their host, and that parasites benefit themselves while harming their host. Van Beneden nonetheless qualified that 'The greater part of those animals which have established themselves on each other, and live together on a good understanding and without injury, are wrongly classed as parasites by the generality of naturalists.' His meticulous observations of parasites doing what parasites do led him to believe that not all parasites harmed their hosts, at least not all of the time. In fact a close reading of van Beneden's 1876 text shows that he was distinctly sensitive to the benevolent side of creatures living in or on other creatures – even those he considered to be true parasites: 'The parasite is he whose profession it is to live at the expense of his neighbour, and whose only employment consists in taking advantage of him, but prudently, so as not to endanger his life.' Although prudence is not a trait one generally associates with parasites, van Beneden was signalling with this word that a parasite aims to avoid harming a host lest it may also harm itself. Understood differently, even if a lowly parasite did not possess prudence, it manifested large quantities of it by any other name.[17]

The keenest observers of parasites therefore felt that if there really were such things as parasites, many of them fostered mutually supportive relationships with their hosts. This could better explain many relationships between creatures, such as the crustaceans, for example, that attach themselves to certain fish or whales so as to clean their hosts by feeding on their cutaneous secretions, and so are tolerated by these hosts. Or there are the crabs and lobsters that nurture worms among their eggs because the worms feed on the unhealthiest eggs and so provide better hatching conditions for those that remain. Or in the case of nematodes infesting roots of desert plants, the plants' cell walls respond by thickening to produce large cavities for better water storage, making life better for both nematode and plant. Examples go on and on: observations since van Beneden's day have revealed that many so-called parasites are providing previously unknown benefits to their hosts. Much like (human) parasites in ancient Greece, scores of biological parasites are being observed providing services while reaping handouts in return. Parasitic humans – the first named parasites – provided entertainment or good conversation in exchange for food or shelter. It makes sense, then, that a witty parasite was often rewarded with a good meal and was simply being compensated by the host for services rendered. Not surprisingly, a host didn't often invite back a dull or morose parasite that reaped more from a relationship then it sowed. Successful parasites pleased their hosts, and some of them worked very hard to please them. Van Beneden saw such behaviour playing itself out in the natural world, where biological parasites employed a variety of ways to please their benefactors, with their benefactors in turn aiming to compensate their parasites: 'The assistance rendered by animals to each other is as varied as that which is found amongst men,' he explained. 'Some receive merely an abode, others nourishment, others again food and shelter.' And if one would observe these creatures carefully, 'it would be but little flattering to them to call all indiscriminately either parasites or messmates.' Van Beneden had looked to the parasites and found many of them to be assisting their hosts.[18]

The dilemma remained, though, that *true* parasites – creatures that took more than they gave – harmed themselves by harming their hosts, and so at some point they would need to cease being parasitical for their own self-preservation. Proposing *prudence* in parasites gave van Beneden a way out of this dilemma by allowing some parasites to continue taking more than giving, but doing so cautiously, modestly and farsightedly. Van Beneden was himself a devoutly religious man, and in his scientific observations he tended to see harmony in all of God's creations, even in parasite–host relations. Only a very few parasites were wholly wicked, he declared; and, like certain noblemen, sought to 'escape, either by cunning, by audacity, or by superior villainy, from social retribution.'[19]

Taking as Well as Giving

In the rise of the science of parasitology, a variety of new terms were proposed for describing relationships between creatures that were meant to refine van Beneden's mutualism, commensalism and parasitism. In the ensuing race of lexicons, one term would pull ahead of the others that denoted proximal co-existence, especially by giving as well as taking: 'symbiosis'. The combination of *sym* and *biosis* ('together living') could contain elements of all three of van Beneden's relationships, though it especially denoted mutual benefit between partners.

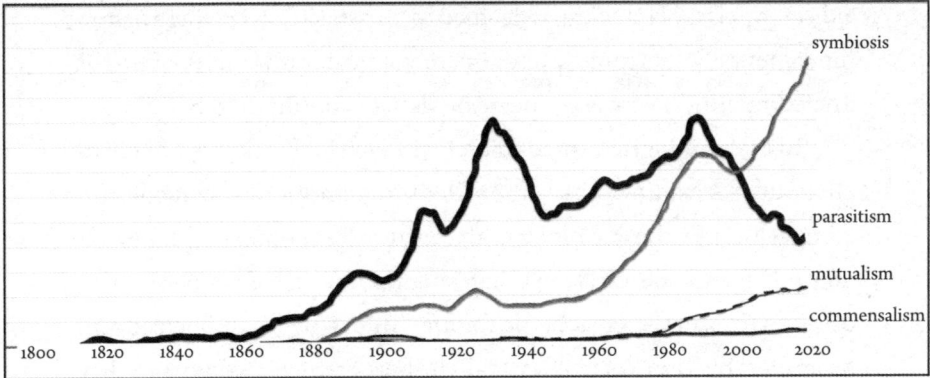

Chart showing the chronology of relative word frequencies
of *symbiosis*, *parasitism*, *mutualism* and *commensalism* that
appear in the literature between 1820 and 2020.

Although some social theorists continued championing a Hobbesian
notion of life that was ruthlessly competitive and selfish, others were
countering that cooperation and collaboration prevailed in the natural
and human worlds. The project of living together often required
accounting for another's needs.

In tracing the long arc of the idea of symbiosis, historian Jan Sapp
has shown how notions of competition on the one hand, and cooperation on the other, were being employed by social activists as well as
natural scientists to support their theories and justify their actions.
Group dynamics of people, animals and plants were coming to be
seen as fundamental to the natural order of things, with religious leaders, politicians and biologists all having a stake in whether those
dynamics were competitive or cooperative. The phenomenon of competition emphasized an individual's needs and abilities, whereas
cooperation stressed group interaction and collective benefit.
Reflecting Karl Marx's preoccupations, a key question was whether
social inequalities stemmed from a tendency to care only about oneself, or whether collaboration and cooperation would ultimately win
the day. Did Darwin's theories imply that there would always be hostility and competition between those with similar needs, or could
evolution by natural selection also favour those who give reciprocal

aid? In the class of individuals deemed to be parasites, were these beings fundamentally antagonistic or ultimately amicable towards their hosts? The competition versus cooperation debate continues to our day.

Russian biologist-turned-anarchist Peter Kropotkin is often singled out as a leading critic of the Darwinian notion of struggle and competition. In the late nineteenth century, Kropotkin began objecting to Darwin's declarations that 'species have tried to overmaster other species in the great battle for life.' This struggle, publicized and promoted by Thomas Huxley and Herbert Spencer, would have key biological and social implications. Indeed, the word 'competition' but not 'cooperation' appears dozens of times in *On the Origin of Species* (1859), and Kropotkin was disturbed that Darwin's book did not allow for much mutual assistance between individuals and species. Historian Daniel Todes reveals how Kropotkin as a young man travelled far and wide across Asia and Siberia. During this Old World counterpart of Darwin's voyage of the *Beagle*, he came to declare that organisms survived extreme environments with help from their fellow organisms. Kropotkin then began extending his observations from the natural world to the human condition, arguing that people also depended on cooperation to survive. Kropotkin's *Mutual Aid: A Factor of Evolution*, appearing in serialized version in the 1890s and in book form in 1902, became a widely read rebuttal to the supposed universal drive of competition. Kropotkin contended that all beings, whether human, animal or microscopic, depended on cooperative acts to survive common enemies, limited resources, temperature extremes and other environmental challenges. Kropotkin argued that population pressures were much less severe than Darwin's theories assumed; rather, life's real challenge was dealing with environmental stress, a challenge alleviated through partnerships with other creatures. Kropotkin noted that Darwin himself would in the *Descent of Man* (1871) acknowledge how 'in numberless animal societies, the struggle between separate individuals for the means of existence disappears, [and] how struggle is replaced by co-operation.' Kropotkin then added that 'even as regards the lower animals we may glean a few facts of well-ascertained co-operation.'[20]

Mutual Aid was not a biological treatise, and was written primarily as a call to political action, meant to reinforce rising anti-capitalist movements that mistrusted a privileged elite who owned the means of production. In the decades before Kropotkin's bestseller, there had been other Russian naturalists extolling cooperative virtues that they (thought they) saw in the field and under the microscope. One of these theorists was Karl Kessler, a zoologist working in St Petersburg who delivered the 1879 lecture 'On the Law of Mutual Aid'. Although Kessler agreed that competitive struggles held sway in some animal interactions, he maintained that mutual aid was the rule throughout the animal kingdom, both within and between species. Individual mammals, birds, fish and insects often helped each other, he asserted, because of their 'need to reproduce [which] leads them to approach and support one another'. Part and parcel of Russia's rising socialist sentiment, Kessler's arguments would prove broadly appealing to his audience, which on one occasion happened to include Kropotkin.[21]

The growing popularity of mutualism was used to explain human as well as natural phenomena and found wide appeal across Europe. Developing alongside Russia's cooperative sentiment was France's movement of Solidarism, which emphasized the role of collective responsibilities in guarding against *laissez-faire* policies. Historian Michael Osborne points out that French naturalists were likewise questioning theories of natural selection that relied solely on competition. For example, Edmond Perrier, director of the Natural History Museum in Paris, advocated a theory of evolution that relied on cooperation as well as association. Raphaël Blanchard would bring the debate about cooperation to the level of parasites. In fact, the medical zoologist Blanchard coined the term *parasitologie* – the science of parasites – while teaching that the creatures of this study field are the scourge of humanity, since they hindered colonization by spreading tropical diseases and generally disrupted the noble project of cooperation and Solidarism.[22]

For naturalists who focused on organisms that lived alongside each other (or 'organismal associations'), lichens had become a favourite study object. Displaying spectacular colours and surviving for decades on rock surfaces, these curious organisms resembled plants but did not act like them. A few enquiring botanists began proposing lichens to be combinations of two or more species. Observations under the microscope lent greater credence to a theory first proposed in 1877 by Leipzig's Albert Frank that lichens were indeed unions of algae and fungi; yet Frank felt uncomfortable describing either alga or fungus as parasites in this union. He proposed instead that both organisms were joined in a reciprocal relationship of *Symbiotismus*: 'We must bring all the cases where two different species live on or in one another under a comprehensive concept which does not consider the role which the two individuals play but is based on the mere co-existence and for which the term symbiosis [German: *Symbiotismus*] is to be recommended.' In subsequent publications by Frank as well as by his more established colleague Anton de Bary, symbiosis gained currency as a neutral association between distinct creatures. Frank and de Bary were therefore unifying van Beneden's tripartite notion of mutualism, commensalism and parasitism into a single concept of symbiosis, proposing it to be the 'living together of unlike named organisms'. But they also felt that in the case of lichens, algae had become so adapted to co-habiting with their fungal partners that they could not survive outside this association – implying that there were elements of co-dependence between the two organisms, even co-advantage.[23]

Following the lichen theories, naturalists began finding symbiotic unions in all corners of life. Frank himself would eventually turn his attention from lichens to bulbous growths on legume roots, proposing that such structures were actually fungi that provided the plant with nutrients in exchange for plant secretions. This relationship Frank called 'Pilzsymbiosis' and in 1885 he christened these fungal associations 'mycorrhiza', a term still in use today.

So even as symbiosis was conceived as a mere physical union of species in the vision of Frank, by the late nineteenth century the term was denoting a relationship of mutual benefit between species. Perhaps as the concept of symbiosis spread from the plant world to the animal world, creatures could no longer be envisioned as merely associating with one another without some sort of mutual feeling. Animals seemed to harbour more purpose than plants, with greater abilities to make choices about how they lived, as well as with what or with whom they associated.

Yet 'symbiosis' would not be applied to human relations for decades, even if naturalists were widely accepting the concept. Instead, 'mutualism' remained the preferred term for denoting cooperative human interactions. Various 'mutual societies' were established in Europe and beyond for providing cooperative logistical and financial assistance to groups of people, especially in the days before the advent of effective welfare states. And parasites, for their part, presented a dilemma to advocates of cooperation. In fact, as 'symbiosis' and 'mutualism' became more popular in biological and social circles, parasites were increasingly assumed to be disruptive forces. Parasites, after all, were viewed as freeloaders that (or, who) depended on their hosts in their habit of receiving more than giving: parasites took for themselves, only worrying about their hosts when their host's viability threatened their own survival. Cooperative relationships seemed the natural order of things until parasitic tendencies crept in to corrupt beneficial unions. By the late nineteenth century, parasitism was the *bête noire* of cooperation.

Kropotkin had himself lamented parasites' selfish behaviour. In the introduction of *Mutual Aid*, 'parasitic growths' were singled out as the source of evil that served to unravel society's 'mutual aid institutions', be they tribe, village community, guild or medieval city. He asserted that a glance at history demonstrated that there were always rebels who 'endeavored to break down the protective institutions of mutual support, with no other intention but to increase their own wealth and their own powers'. In Kropotkin's earlier and more radical

The Conquest of Bread (1892), published when he was working on the first passages of *Mutual Aid*, the parasite is more severely chastised. Ever more wealth is being concentrated in the hands of the few, Kropotkin lamented; 'the swarm of parasites is ever increasing.' Waves of followers would read and praise *Mutual Aid*, using Kropotkin's celebration of cooperation to support their own socialist views while locating more instances of it in nature and society. Italian leftist revolutionaries were still lauding and republishing the book in the 1920s, pleading for grassroots collaboration to overthrow the owners of production.[24]

One naturalist who emphasized cooperative behaviour was Hermann Reinheimer, who cited Kropotkin in his *Evolution by Co-Operation* (1913), a popular work suggesting that organisms had greater chances of survival if they maintained reciprocally supportive partnerships, particularly ones in which each partner sought to outdo the other in generosity. Here again, a crucial hindrance to cooperation was parasitism: Reinheimer explained that a host's reproductive capacity was ordinarily high until 'the parasite had started on its career of prodigality'. Cooperative behaviour was treated more methodically in Reinheimer's next work, *Symbiogenesis* (1915), which he described as a 'physiological partnership between individuals of different species' applicable 'to the wider bio-economic form of co-operation which underlies evolution and unites all organisms in a vast web of life'. Here, the meaning of symbiosis was moving beyond a relationship of simple association to one of active cooperation between members, be they human or non-human.[25]

Zoologist Paul Portier, working in France, would take symbiosis to still wider applications by speculating, for example, that many bacteria living in and on larger creatures often benefited their hosts, rather than simply inflicting disease and dysfunction. Certain organelles found within eukaryotic cells, Portier conjectured, had themselves once been free-living bacteria that entered into and contributed to the functioning of a larger cell. In his amazingly clairvoyant *Les Symbionts* (1918), Portier proclaimed that sub-cellular structures such as mitochondria are actually bacteria that provide vital functions to their cellular host. This controversial theory of beneficial micro-organisms moving into other organisms turned parasitism on its head by suggesting that a body's co-inhabitants played critical roles in nutrition, development, heredity and even species origins. Although such revolutionary ideas were largely dismissed in Portier's day (and his book was largely forgotten), this notion of organisms joining together to create super-organisms would be reborn under various guises in coming decades. 'Symbiogenesis', a term coined by the Russian Konstantin Mereschkowski in 1910 to describe the incorporation of symbionts, would be recycled by Reinheimer, developed by Portier and revived and then trumpeted by such theorists as Lynn Margulis in the 1980s. For microbiologist Margulis and her followers, the joining of organisms through

Parasitic mistletoe (*Viscum album*) growing on a linden tree (*Tilia cordata*) near Munich. According to biologist Walter Koenig and his team at Cornell University, a tree that hosts abundant mistletoe may well be a sign of its robust health rather than an indication of its decline.

symbiogenesis is one of the crucial mechanisms by which creatures evolve to produce new organisms and multiply the Earth's biodiversity. Life flowered into its multitudes through the action of cooperative symbionts, which earlier theorists had assumed to be despicable parasites.[26]

Which Came First, a Parasite or a Mutualist?

A simmering question for these early observers was whether parasitism generally led to mutualism, or whether mutualism led to parasitism. That is, were parasites defectors in a symbiotic and mutualistic relationship, or were parasites a preliminary stage in becoming mutualists? More specifically, did selfishness beget cooperation, or was it the other way around? Such group dynamics had wide implications for explaining the past and predicting the future of biological as well as social systems. Continuing to our own day, this chicken-and-egg question centres on whether selfishness leads to or arises from selflessness.

In the views of social theorists Herbert Spencer and then Peter Kropotkin, selfish individuals tended to prey upon cooperative unions, with tribes, villages and guilds being readily corrupted by parasitic tendencies. Naturalist Hermann Reinheimer likewise viewed parasitism as a disruptor of cooperative symbiosis: 'Parasitism is a failure because it violates what I conceive to be the fundamental law of co-operation in the biological world.' There is 'frequent degeneration of recognized cases of symbiosis into parasitism'. Similarly, Maurice Hall, the carbon tetrachloride advocate, viewed parasites of domestic animals and of food crops as purveyors of evil, with the only rational response being to eradicate them.[27]

But other theorists countered that parasites, far from simply inflicting harm, could foster harmonious unions. Van Beneden had early on claimed that a typical parasite acted prudently so as to not inflict too much harm on its host. For the likes of Portier, cellular organelles were initially free-living creatures that had parasitized cells before learning to cooperate with them. Even George Nuttall, a Cambridge parasitologist between the world wars, found it 'difficult to imagine that symbiosis originated otherwise than through a preliminary stage of parasitism'. Some of the strongest support for this parasitism-begets-mutualism doctrine came from Theobald Smith, a leading American microbiologist and contemporary of Nuttall. Smith felt that a parasite living in a host relies on offensive as well as defensive strategies: 'one for active injury, the other for more passive self-protection'. Smith believed that the host's as well as the parasite's interests were best served if each learned to tolerate the other. Although a parasite's initial entry into a host typically causes disease and disruption, said Smith, such ill effects became less harmful over evolutionary time: 'we are gradually approaching the view that the relation between host and the invasive parasite is a highly complicated give-and-take process.' Regarding malaria parasites, Smith pointed out that they sometimes did not cause disease at all: indeed, large numbers of human *Plasmodium* carriers are immune or mostly so, since they are capable of 'transmitting the infection when on the surface no malaria was in sight'. Smith and his followers would go on to develop the so-called Law of Declining Virulence, which asserted that parasites become less and less noxious to their hosts as time went on. Pathogenic parasites would eventually cease inflicting pain and suffering. The far-ranging implication here was that humanity's troubling parasitic diseases would fade and finally disappear.[28]

So while Reinheimer considered parasitism to be 'an extreme instance of aberration' whereby our internal creatures aimed to disrupt once-harmonious mutualisms, Smith singled out the 'aberrant parasite' as the source of disease due to its initial maladjustment to its host. Smith felt that parasites would eventually settle into a reciprocally beneficial relationship with their hosts. Or as one of Smith's colleagues, E. M. Freeman, expounded in 1937, 'symbiosis may be regarded as mutually beneficial partnerships arising from and due to parasitism.' Hold that carbon tetrachloride: here was strong rationale for discontinuing the quest to kill every last parasite.[29]

In coming decades, the scientific community would continue to debate the outcome of host–parasite relationships, with pessimists predicting heightening infectious disease as parasites continued antagonizing hosts – including humans – but with optimists contending that diseases would fade as parasites and hosts became accustomed to each other. Noted geneticist Theodosius Dobzhansky would side with the optimists, claiming that parasitic tendencies are unstable and 'tend to disappear and be replaced by cooperation and mutualism'. The case of exploding rabbit populations in Australia that were checked by myxoma virus led greater credence to a parasite's diminishing virulence when it was observed that this introduced virus became considerably less lethal over time. As recently as 1993, Nobel laureate Joshua Lederberg still asserted that host and parasite coevolve to lessen the harm that one inflicted on the other: 'Natural selection, in the long run, favours host resistance, on the one hand, and temperate[s] virulence and immunogenic masking on the parasite's part, on the other.' One wonders if those on the front lines of the battle against malaria also mulled over such claims, or ever doubted the assumption that the world would be a better place if it became *Plasmodium* free.[30]

Some of the latest experimental evidence suggests that a parasite's virulence rises or falls depending on the relationship with its host, the timing of infection and the health of each member. Joseph Schall presents mathematical models that simulate parasite–host relationships to reflect cooperation as well as competition, thereby indicating 'some of the most intriguing findings in all of ecology and evolutionary biology'. To judge how far parasitism will lead to mutualism, Schall pleads for more data: 'Data on the actual costs of parasitism . . . data on costs of infection . . . data on a greater variety of taxa'. Judging whether malaria's *Plasmodium* may be adapting to its human host therefore requires us to seek more information. We turn our attention in the next chapter to the unexpected ways that parasites affect hosts, some of which are not only adaptive but beneficial.[31]

Cooperative Parasites

In wondering what else might be gleaned from the history of parasitology, we can consider the case of tapeworms. In sifting through the old medical literature, one finds that human tapeworm infestation was often so severe that health providers measured these worms by length rather than number. According to a revealing report from 1867, a certain British subject treated with 'extract of kamala' went on to pass 'three yards of *T. saginata*, and then on being treated again almost immediately with oil of male fern, passed a further 11 yards together with the head, making 14 yards of tapeworm in total'. Just as gut-wrenching is an 1893 report from Marseille which reported that of 2,686 patients examined, some 87 per cent of them carried one tapeworm, 7.8 per cent carried two, and 2.3 per cent hosted three or more. It also seems that proglottids – tapeworm segments – had an annoying tendency in those days to drop out at inopportune moments:

> The segments pass when the patient is standing quietly and falling into his trousers, he suddenly has a moist and cool feeling about the legs, and when he seeks to free himself from this unpleasant sensation, finds a single proglottis attached to or creeping about the leg. Women especially are afraid lest the proglottides should fall unperceived upon the ground when they are walking or standing.[32]

Although few of these displeasures arising from tapeworm infection were mentioned in earlier, medieval writings, David Grove notes that by the seventeenth century, 'all manner of ills were ascribed to this worm,' bringing on a search for more effective and less nauseating anthelmintics. But the actual negative health effects of tapeworms remained hard to document. A methodical 1937 helminth investigation of a Russian village revealed that very few of the inhabitants showed adverse clinical effects, even though almost 90 per cent of them were

Nine-year-old Brazilian boy displaying the hookworms
and tapeworms he expelled, *c.* 1920.

found to be carrying tapeworms. Similarly in Finland during these
years, where tapeworms were likewise observed to be ubiquitous
inhabitants of the human gut, a study reported that just 0.1 to 0.5 per
cent of infected persons suffered any kind of anaemia from them. In
fact, medical investigators in these later years found it difficult to

ascribe any serious human ills to this sinuous co-inhabitant. We can then ask: is it possible that these tapeworms were living peacefully within their hosts, even cooperating or collaborating with them to make life better for both?[33]

Certainly when acknowledging the many creatures living within our own bowels, it can be a stretch of imagination to suggest that anything like 'cooperation' can exist between us – as if a tapeworm could consciously choose to work with us to complete a common goal. If one reads through the many investigations of cooperation between or within species, or about any evolutionary development of cooperation, one easily discerns ideology colouring observation. Mutualism and mutualistic symbioses are relationships in which all benefit, with the observer tending to assign human traits to these relationships. Pierre van Beneden was convinced that many of the bodily creatures he observed in his specimens were acting prudently and perhaps selflessly when interacting with their hosts. Peter Kropotkin taught that cooperation is a universal attribute, driving animal relationships as well as human societies. Both thinkers effortlessly assigned human characteristics to a non-human world, even if cooperative tendencies may be rather different between animals and humans.

Between the world wars, animal ecologist Warder Allee also called attention to 'vague, unconscious mutual co-operation' between associations of non-human organisms, whether in the laboratory or in the field. Allee was convinced that partners in animal groups often sought to provide benefits to each other – a phenomenon that Allee referred to as 'proto-cooperation' – even if his sceptics reminded him that the mere existence of animal associations did not necessarily imply cooperation. In arguing for the cooperative nature of parasites, Allee approvingly cited Theobald Smith, asserting that 'successful parasitism may be regarded as a compromise or partial truce between two living populations.' When pathogenic parasitism clearly indicated antagonism rather than cooperation between parasite and host, Allee echoed Smith in explaining that occasionally 'the truce may be broken and severe injury result for either parasite or host.' Various science popularizers and political theorists continued reiterating Allee's points, as when political philosopher Radhakamal Mukerjee taught that 'Bioecologic co-operation or, to use another term, symbiosis, organic and social, is the key to the permanence of man's civilization, his works and experiences on the earth.' Mukerjee was part of a long string of advocates of cooperation, whether or not parasites have actually been cooperating.[34]

At the 1980 parasitology conference in New Orleans, Paul Beaver did not mention that a parasite was *cooperating* with its host; yet he also felt that most parasites were not pathogenic. 'Patients generally are eager to rid themselves of their parasites even when they are essentially harmless – or even when they are nonexistent,' noted Beaver; parasitologists 'must expect occasionally to encounter people whose most troublesome parasites are those that exist only *in the mind*'. As we explore in the next chapters whether we should be fearful of our parasites, we might at least agree that cooperation is the project of two or more parties working together to achieve a common purpose. Even though it is hard to conceive of lowly parasites as upholding their end of this complicated affair with human beings, they have indeed survived with us, in us, since before hominids began marching across continents. As it seems that two creatures living together cannot help but affect one another, whether for good or for ill, it follows that a parasite and a host who live together must be modifying one another's lives. It then seems reasonable to believe that some of our parasites, some of the time, improve their lives by improving our lives.[35]

3

The Benefits of Being Parasitized

In China there existed from ancient times the widespread
belief that one should have at least three worms in order
to remain in good health.

REINHARD HOEPPLI, *Parasites and Parasitic Infections in Early*
Medicine and Science (1959)

'Coprolite' is a palaeontologist's euphemism for fossilized faeces.
In the 1960s Gary Fry, a doctoral student sifting through sediments at Danger Cave in North America's western desert, found a variety of human coprolites that would be carbon dated to 10,000 years ago. When turning a microscope on his specimens, Fry revealed a miniature world teeming with pinworm eggs solidified in time. As his 1969 *Science* article recounts, the coprolites he unearthed at various depths in the cave revealed traces of human intestinal worms that proved to be some of the oldest ever found in the Americas. Fry was following in the footsteps of Eric Callen, the 'father' of human coprolitology who had studied the contents of petrified faeces extracted from much older Neanderthal sites in France. Once discarded as archaeological clutter or sometimes launched as flying discs by fatigued excavators needing comic relief, these scatologic treasures provided new vigour to the young field of palaeoparasitology. Here was tangible proof that people have long been nourishing intimate relationships with their parasites.[1]

In our current world, it is estimated that more than a billion humans continue to host pinworms in wealthy as well as in poor countries. Yet almost none of these human hosts suffer serious consequences from *Enterobius vermicularis,* the centimetre-long worm that varies in numbers from person to person, from place to time, from gender to class. During the Second World War, residents of Guam witnessed a pinworm infection rate of just 1 per cent, whereas in Amsterdam at the same time, almost every individual was found to be carrying this nematode. Pinworm infection rates also vary by age, with children much more likely than adults to test positive for this worm; a recent study found that half of British children under age ten play host to it. The latest findings of the Human Microbiome Project further reveal that our parasitic load and diversity depends on bodily health, as well as on season, diet, pre-existing microbiota, occupation and exposure to other hosts, among many other factors. Like all ecosystems, each of our bodies is a constantly adjusting network of biotic and abiotic elements such that slight chemical or temperature shifts within our tissues can produce dramatic fluxes in micro-organismal composition.[2]

Danger Cave, at the edge of the Great Salt Lake in North America, where 10,000-year-old human coprolites have been excavated to reveal traces of pinworms.

To consider another example from deep time, Ötzi the Iceman, who wandered the high Alps 5,300 years ago to be pierced by an arrow and then thrown into a crevasse, was also a walking microbiome. Although pinworm eggs have been hard to detect in Europe's oldest mummy, now sitting in an Italian museum, he is loaded with traces of whipworm. In fact, Ötzi may have experienced a good deal of discomfort from the whipworm that lived in his gut. *Trichuris trichiura* typically produces diarrhoea and anaemia in its carriers and may lead to long-term physical and mental harm that is sufficiently serious to convince today's health experts to single out whipworm infection as one of the seven 'neglected' diseases now being propagated by some 800 million carriers worldwide, or one-ninth of humanity. A careful inventory of Ötzi's belongings brings speculation that chunks of birch mushroom (*Fomitopsis betulina*) found attached to a leather lace that he carried in his girdle pouch may have been his personal anthelminthic that he consumed piecemeal for controlling his whipworm infection. It seems that Ötzi, a man believed to be in his forties with Mediterranean origins, suffered like many of us from occasional irksome parasites while simply going on with his life, largely oblivious to the rest of his bodily spectrum of microflora and microfauna. Ongoing inspections of Ötzi and his tissues also reveal that he was a carrier of *Helicobacter pylori*, a bacterium considered by some as an important cause of ulcers, but which other researchers believe is actually a crucial gut inhabitant, useful for slowing the onset of such issues as arthritis and asthma. More curious may be that Ötzi was found to be free from other parasites that are now relatively common, such as *Giardia* and *Cryptosporidium*, which typically play havoc with our modern gastro-intestinal tracts. Ötzi's microbiota were apparently marching him towards ill health at the same time as they were fostering his well-being.[3]

These and other palaeoparasitological observations are demonstrating that our ancestors across the globe teemed with assorted arthropods, protozoa, bacteria and viruses while being strangely free from other bodily creatures that are now widespread. One may wonder to what extent Ötzi's load of whipworms provided him with possible advantages of which we are unaware. Allergologists today are quick to point out that a diverse load of intestinal worms and other gut microbiota may mitigate not only arthritis and asthma, but incursions of fever, various skin disorders, Crohn's disease and other auto-immune responses such as lupus and multiple sclerosis. Carefully administered faecal transplants, by transferring the microbial contents of one person's lower intestine to that of another, can offer remarkable alleviations of these disorders. Although Ötzi was markedly thin and malnourished while crossing the mountain passes that fateful day, we can wonder whether he was largely free from most of our own auto-immune problems. We can also ask how other Icemen and Icewomen, who lived in their cultural world five millennia ago, were affected by their own bodily fauna of commensals, mutualists and parasites. Or, closer to the present, we can wonder how the lives and lifestyles of our grandparents or their grandparents – with their wars, technologies, hopes and biases – were modified by the creatures living inside them, beyond acting as simple agents of illness, suffering and death.[4]

This chapter explores a different side of parasites in human history by recognizing that these creatures have been not only destructive and deleterious, but advantageous and beneficial in various ways, for individuals as well as for communities. Even though the great transfer of microbes and parasites across continents and oceans, for infecting and devastating distant populations, is one of the canons of our understanding of migrating humans, we need to be on the lookout for *non*-lethal, and even beneficial, effects of humans carrying, promulgating and transmitting the plethora of bodily creatures they carry. As observers since George Perkins Marsh made clear more than a century and a half ago, humans have never travelled alone, since their

> Plants and animals often carry their parasites with them, and the traffic of commercial countries, which exchange their products with every zone and every stage of social existence, cannot fail to

transfer in both directions the minute organisms that are, in one way or another, associated with almost every object important to the material interests of man.

As historians Alfred Crosby and William McNeill have since clarified, when explorers and colonizers journeyed around the world and encountered new peoples, they exchanged portmanteaus of creatures, with effects that were as transformative to landscapes as they were devastating to indigenous peoples. These virgin soil epidemics, as Crosby called them, inflicted destruction and death on massive scales in the New World.[5]

Yet one can now begin to realize that Columbus and his crew transmitted more than just disease and suffering when they landed on distant shores. The migrating superorganisms that are *Homo sapiens* carried a spectrum of microbes, in themselves and in their domestic animals and plants, to set in motion an array of consequences that involved more than ill health. Spanish sailors and Caribbean peoples exchanged their symbionts with each other, whereby hosts and parasites joined in relationships of cooperation and not just antagonism. As David S. Jones cautions, we must be wary of accepting simplistic theses of historical epidemics whereby New World peoples easily and quickly succumbed to Old World microbes, for there was much more happening than unequal susceptibilities to these creatures. We need to explore how parasites, and not just pathogens, have been crucial actors in the human story – with some of these parasites acting as mutualists. In tracking the microbial consequences of the world's colonial encounters, one can find mutualistic exchanges that were occurring alongside pathogenic exchanges. In fact, one can begin to appreciate that some organismal exchanges were mutualistic precisely because they were pathogenic, since the hosts often benefited from the horrors their parasites could inflict.[6]

The Horrors of Pathogenic Parasites

In 1907 essayist W.H.S. Jones speculated that malaria may well have been the primary factor leading to the downfall of the entirety of ancient Greek civilization. Jones pointed to archival evidence that malaria was introduced to the Greek isles in the fifth century BC, corresponding closely to when that civilization was tottering towards decline, so that 'it is quite possible for the disease, running a practically unchecked course, to have produced the profound deterioration which occurred in the Greek character during the next century and a half.' To give credence to this epidemiological view of history, one can mention any number of key figures who perished from this parasitic disease. There is evidence that Alexander the Great's death throes in 323 BC are consistent with the symptoms of malaria. Similarly, Pope Innocent III, Marco Polo, Oliver Cromwell, Henry VIII, Mary, Queen of Scots, Zog I, George Washington, Ulysses Grant and Mahatma Gandhi are a few names on a long list of influential leaders who may well have been victims in part or in full of the malaria parasite. Other studies offer up lesser-known authors, scientists, artists, leaders and musicians who likely succumbed to this notorious disease, reflecting *Homo sapiens'* chilling experiences with one of its longest cohabitants. Palaeopathological evidence suggests that malaria parasites first appeared in tropical climes, to eventually be exported to Europe and the rest of the world, often in the veins of enslaved human cargo. A recent study relying on DNA sequencing of *Plasmodium* genomes concludes that *Plasmodium vivax* is the world's most widespread malaria parasite, being the first to arrive in the Americas, probably through explorers and merchants, followed by the more deadly *P. falciparum* that arrived a century later with the first African slaves. Viewing malaria parasites as just one species of many deadly creatures that infect human beings, it is easy to appreciate why a dizzyingly large fraction of the world's human population has fallen from the pathogenic effects of our bodily creatures, small or large, imported or exported, insect-transmitted or not – from packets of viral DNA to bacteria, and from protozoa to helminths and beyond. Although the degenerative diseases of old age, heart attacks and cancers kill more people today than infectious, communicable diseases of microbes, such was not the case before modern medicine was able to rein in many of the lethal effects of our everyday pathogens. A tabulation of causes of death in medieval England shows that most of these are attributable to pathogenic diseases, ranging from 'consumption' (typically tuberculosis) to 'rising of the lights' (pneumonia) and 'ague' (usually malaria), in combination with spikes of other maladies including plague.[7]

To appreciate the waves of suffering that Crosby and McNeill trace to the transmission of pathogens, one can review what's known about the historical demography of the Americas, which with each new study generally sees numbers of indigenous victims marching upwards. William Denevan recounts, for example, that the population of native South, Central and North America, which was initially estimated to be in the tens of millions at European contact – plummeting to some 1 million individuals by 1900 – has with more research been pushed to 50 million and higher, with the Aztecs' main city of Tenochtitlán now considered to have housed more people than Paris when Columbus first crossed the ocean. Subsequent indigenous die-off rates of 90 per cent are considered conservative in the latest historical revisions of the Western Hemisphere's population changes. Yet the project of revealing the exact proportions of the New World's human destruction may be less critical than simply acknowledging that untold numbers of people lost their lives in the decades and centuries after the Spanish first arrived on Caribbean shores, meaning that whole cities, indeed 'entire ways of life', as Charles Mann has written, 'hissed away like steam'.[8]

Historical epidemiologists believe that the deadliest illnesses of this massive microbial exchange did not initially include malaria, but rather smallpox, measles and influenza, followed by mumps, measles, typhoid and tuberculosis, which all stemmed from noxious bacteria and viruses, and were all transmissible by a cough, a handshake or a blanket. Then, in the wake of the diseases of direct contact, subsequent microbes responsible for yellow fever, typhus and malaria depended on insect transmitters to move from person to person and spread their misery. Of all these iconic diseases, perhaps only malaria is the subject of classical parasitology, since here, the dangerous organism being transmitted from host to host is a relatively large protozoan (or eukaryote). But the other disease-causing microbes might also be considered parasitic since, even if bacterial or viral, they also live on or in foreign bodies and require these organisms and their resources to survive and reproduce before moving on to their next host. The tale of how this collection of culpable pathogens was transmitted over the centuries from colonizer to colonized to obliterate indigenous populations has become one of the acknowledged cataclysms of the world's biological exchange. Experiences in all isolated parts of the world, from New Zealand and Australia to Greenland and Siberia and nearly every remote land in between, tend to follow this frightful storyline of contact from afar followed by transmission of foreign pathogens leading to dramatic diffusion of disease and local demographic collapse.[9]

Yet by shifting paradigms, one realizes that this microbiological exchange can be viewed as beneficial and advantageous and not simply antagonistic and destructive. The many deadly and debilitating viruses, bacteria and protozoa that passed from person to person and from group to group can be seen in different perspective if we take the viewpoint not of the colonized but of the colonizer. The recipient of virulent microbes sees only the threat of illness. But the transmitter of such microbes would see such creatures as providing critical advantages, since these could inflict disease on the colonized to facilitate domination. A bodily creature that is a competitor for one is a cooperator for another, providing its host with beneficial services, even working with its host to further the interests of both. How else could a motley collection of soldiers led by Hernán Cortés hope to have any success invading the city of Tenochtitlán to threaten the Aztec empire? From the perspective of Cortés, the smallpox and measles that he and his men

Preventative vaccinations given during the Russian anti-plague expedition
to Anzob in Russian Turkestan (subsequently northwestern Tajikistan), 1898.

carried – even if unknown to them and invisibly multiplied and transmitted to their enemies – was the best military strategy conceivable. Here
were pathogenic partners of immense utility to their hosts. Toribio
Motolinía, the Spanish monk who took up residence near the Aztec
capital at the time, famously wrote of the smallpox epidemic spreading
before his eyes that 'more than half the population died . . . They died
in heaps, like bedbugs.' The Aztecs' loss was the Spaniards' gain, all mediated by the Old World's dangerous microbes. The staggering demographic declines do not change, but the perspective of whether such
population crashes were indigenous losses or European gains depends
on who is telling the story, and whether bodily creatures are adversaries
or allies. Even if it is certainly a heartless gesture to label the Old World's
conquest of the New as a project of microbial alliance, this victory
surely reveals one of the supreme benefits of being parasitized.[10]

An ecologist might propose that Europeans benefited by 'enemy
release' when they landed in these globally isolated places and dumped

their own debilitating parasites onto their competitors, who generally harboured much less resistance to such parasites. An epidemiologist would add that the immunological efficacy of Columbian-era Amerindians against many Old World pathogens had progressively declined in the 13,000 to 33,000 years since their ancestors had passed across the Bering land-bridge's bottleneck. In that intervening time, New World and Old World peoples alike continued to develop disease resistances to their own proximate pathogens, but such pathogens became isolated in their separate worlds, diverging over time in type and virulence. By 1492 New World physiologies were not prepared for the onslaught of supercharged Eurasian and African microbiota and their diseases. In some cases, seasonality and impoverishment heightened Amerindians' susceptibility to noxious microbes. Yet in the immediate consequences of the Columbian exchange and the American demographic collapse that ensued, invading Europeans – along with Africans coerced to join them – demonstrated how fortunate they were to have brought along their most dangerous bodily creatures. For the Amerindians' own part, the most important microbe they brought to the encounter may have been the syphilis bacteria, even if it is still unclear whether this pathogen originated in the Old or else New Worlds. Still, if syphilis is indeed an American gift to the conquerors, then it provided a bit of compensation for all the misery unleashed in the other direction.[11]

Of course, the Old World's subjugation of New World peoples certainly involved much more than microbes, for there were important cultural contrasts between and within these peoples, including differences in weapons, trade, social practices and spiritual traditions, among many other factors, as well as different landscapes, climates and disease carriers, that may all help explain why certain indigenous groups were more susceptible or more resistant to European invaders than others. But in considering Europe's and Africa's and Asia's larger expansions across the continents and oceans, one must acknowledge the crucial roles played by the differential immunities of human hosts to their parasites. Since Old World invaders' immediate ancestors had been dealing with a larger pool of virulent and transmissible microbes, while propelling their potency with rising human densities and accelerating global trade, their more virulent microbes ended up providing them with brutal advantages in their encounter with many New World peoples. The story of the devastating effects of contagions transported from Asia to Europe, and from Europe and Africa to the Americas and the Pacific, is now retold in every textbook of world history. But devastation for one meant triumph for the other, with such stories depending on whether one takes the perspective of conqueror or conquered.

The Benefits of Pathogenic Parasites

Beyond a parasite's ability to inflict disease on their host's competitors and enemies, there remain a variety of more subtle and indirect ways by which parasites can benefit their hosts. Historian John McNeill emphasizes the role of the temporal lag of being parasitized, since longer exposure to a dangerous microbe often gave the host and its descendants greater acquired immunities against it. This immunity lag is central to his explanations of how the Caribbean's first European colonists were able to conquer various favourable lands, and then how their descendants were in turn able to fend off newly arriving settlers and aggressors. In this way, smallpox and measles initially imported from Europe to Hispaniola, Barbados and Cuba served to disrupt and decimate indigenous societies living there, facilitating Spanish conquest. Then a hundred or so years later, by the eighteenth century, with Spanish and African settlers established on these isles, their acquired resistances to the next wave of diseases of malaria and yellow fever helped defend them against subsequent invaders from England, France and the United States, who were much more susceptible to these latest diseases. John McNeill's narrative fine-tunes Crosby's 'Columbian Exchange' by giving the Old World's first colonizers a one-two punch advantage in the New World.[12]

But instead of telling this three-hundred-year tale as one of disease susceptibility of colonized followed by superior immunity of colonizer, Caribbean history can be recast as a broader story about the benefits of being parasitized. Spanish conquistadors benefited from their initial composition of parasites, and two centuries later, their descendants were benefiting from their newly contracted parasites. Such benefits hold true for understanding Africa's colonial history as well, since indigenous Africans with heightened immunities to their own parasites were able to unleash them on the first European colonizers disembarking in Africa. A Nigerian's defence strategy of cooperating with one's parasites and microbes often started in the womb, to be further promoted through mother's milk and early childhood. Malaria parasites along with yellow fever viruses, for example, were continually being transmitted between local Nigerians via mosquitoes, so that these pathogens-turned-allies presented a formidable barrier to British merchants and soldiers sweeping into Nigerian ports to extract whatever useful cargo they could find. Mosquitoes were thus accomplices with the microbes for helping coastal Nigerians fend off foreigners. Philip Curtin reports that during a British steamer expedition up the Niger River in 1841, four-fifths of the crews caught malaria, with a third of them soon succumbing to the disease. Little surprise that today one resident of Ibadan, Nigeria, gives special praise to the mosquito for helping protect his homeland against these early foreign incursions: 'Let us give thanks therefore to that little insect, the mosquito, which has saved the land of our fathers for us . . . The least we can do is engrave its picture on our National Flag.' In a similar line of reasoning, McNeill reiterates the role of the mosquito in aiding colonial Americans to defend themselves against their British oppressors by pointing out 'How the Lowly Mosquito Helped America Win Independence', with the scourge of malaria being the secret weapon that Americans deployed against their enemies after developing immunities against it. Yet one can realize that for both Nigerian and American patriots, thanks might be given, perhaps less to mosquitoes, and more to the wriggling microbes that such mosquitoes carried and transmitted – for it was these microscopic organisms that ultimately wreaked such havoc on human aggressors. Mosquitoes merely transported the sickening little protozoa and, as it turns out, provided an insect gut for completing their life cycles.[13]

Timing and location of microbial exposure are therefore key to reaping the full benefits of one's pathogenic parasites. Recent blood tests show that in rural Nigeria today, three-quarters of a sample of villagers carry plasmodia in their veins, even if very few of them actually suffer from any serious ill-effects of malaria. As was true during the days of Britain's early attempts at exploiting Nigeria, *Anopheles* mosquitoes today are still transferring malaria parasites between human hosts so that, through acquired and even innate immunities, Nigerian physiologies have been able to put up a good deal of resistance to the parasites' ill effects. Acquired immunity takes months or years, typically becoming more effective if stretched out over a lifetime to allow parasite and host to become better accustomed to living with one another. Innate immunity results from the natural selection of favourable adaptations to parasites, which are encoded on DNA and then passed from generation to generation. Although precise biochemical mechanisms of acquired and innate immunities are still being worked out, it has been shown that extended cohabitation of parasite and host often diminishes the negative effects that each inflicts on the other. So once New World peoples were exposed to the initial shock of Old World parasites, survivors would also begin building up their own immunities against them. As Crosby explains, 'Indian contact with Europeans and Africans seemed to lead not to the total destruction of the Indians, but only to a sharp diminution of numbers, which was then followed by renewed population growth among the aborigines.' Historian Paul Sutter calls attention to such trends in the brutal struggles with malaria and especially yellow fever during the building of the Panama Canal, with the French finally giving up by the 1890s and the United States eventually prevailing two decades later, but with resident Panamanians at this time harbouring such resistance to these illnesses that they showed little patience for the battery of invasive physical and chemical measures aimed at controlling these diseases. As sanitary inspectors noted at the time, 'being mostly immune, [local residents] took small interest in yellow fever eradication.'[14]

Only when there have been long lapses of parasite exposure does a person typically become seriously ill upon receiving a fresh exposure to parasites. James Webb Jr recounts the tale of how a prolonged drought in Ceylon during the 1930s dried up most mosquito habitat there, serving to decimate numbers of this insect across the island, and consequently the malaria that it carried – but with the result that when the rains returned several years later, so did malaria, striking with heightened ferocity in a population that had lost much of its acquired immunity to the disease. Such experience also suggests, for example, that there may be drawbacks to a current public health strategy of relying on wide distribution of bed nets in malarial areas, since by erecting barriers to mosquitoes and preventing people from ever developing any resistance to the disease, they may at some future time become seriously ill if they ever fail to put up their bed nets or are otherwise exposed to the parasite. Randall Packard corroborates this warning by pointing to a study in Tanzania in which 'children protected by nets would have a reduced level of immunity in later life.' By the same reasoning, temperate malaria carriers exposed only seasonally to *Plasmodium* may suffer more severe symptoms than do their tropical counterparts who are exposed continuously to the parasite. As a group of epidemiologists put it, 'Routine exposure to hyper- to holoendemic malaria protects a majority of individuals while killing a minority.' Such insights mean that when malaria was still common in Sardinia a century ago, it may have played a greater role in the lives and lifestyles of these islanders than it did in those of people living in tropical Africa.[15]

One can therefore say that humans typically reap more benefits from their parasites if they are continually, rather than periodically, infected by them. In the case of smallpox, Crosby notes that 'populations untouched by smallpox for generations tend to resist the disease less successfully than those populations in at least occasional contact with it.' Not surprisingly, perhaps, Napoleon's most recent army recruits who moved from the countryside to Paris were often the most susceptible to infectious diseases, since their bodies had only rarely been exposed to the full range of threatening microbes that urban dwellers were continuously exposed to. Still today, rural folks travelling for the first time to big cities may suffer inordinately from the ill effects of parasite exposure, since their bodies have been given fewer chances to become accustomed to these intimate creatures and the immunological benefits they can offer.[16]

One also realizes that to reap the fullest benefits from one's parasites, it may be best to begin hosting them at a very young age, even before birth. A lifelong relationship between host and parasite can mean both organisms learn to cooperate from the earliest stages, with each side working to resolve hostilities early on. Although humanity's childhood infectious killers are notoriously tragic, with the Grim Reaper still harvesting frighteningly high numbers of newborns and toddlers around the world, the prospect of contracting these same diseases later in life can prove even more devastating. Infants typically develop a range of disease immunities and resistances from their mothers through the placenta and breast milk. Thus polio, flu and even yellow fever may be debilitating to an infant but fatal to an adult: better to suffer from these microbes when young than die from them when middle-aged. One of the early challenges of polio vaccination programmes in the 1950s and '60s was that vaccinated adults were no longer providing their young children with the immunological signals of polio, thereby making these youth more susceptible to the disease if they entered adulthood unvaccinated and were then exposed to the virus. One gains the most from one's pathogenic parasites when releasing them on enemies who are not only immunologically less adapted, but who are also full adults with little experience at making peace with these creatures. Such was the situation of the first Spanish arriving in the Americas, who had grown up with smallpox and measles before unleashing them on highly susceptible adult indigenous populations. Descendants of the Spanish Americans who were exposed to and survived imported malaria and yellow fever then presented formidable barriers to the next wave of Northern Europeans arriving without such exposures. These late-arriving colonists to the Caribbean encountered, not a New England, but a New Nigeria (or more accurately, a New Yorubaland). By the eighteenth century Spanish Americans living in these tropical isles were reaping the benefits of being parasitized by African parasites transported there earlier, which were defending their hosts accordingly.[17]

As mentioned, living with parasites over several generations typically means that it is the most resistant of hosts who thrive and survive and then pass these resistances on to their progeny. In this way, innate resistances and genetic immunities are born and passed on to the next generations, even expanded to a whole people and continent. The well-known sickle cell trait is a genetic adaptation to a malaria-saturated environment in which sickle-shaped red blood cells can provide greater advantages to the host than normal blood cells: even if sickle cells carry less oxygen for the host, they also limit the infection of *Plasmodium*, since the parasite relies on healthy blood cells to develop. As a result, people who are neither sickle-cell free nor sickle-cell saturated (homozygous), but sickle-cell heterozygous – carrying some but not all sickle-shaped blood cells – can best accommodate their demanding parasites, which then offer compensation by debilitating their host's competitors and enemies who lack such sickle cells. A person living in a malarial environment therefore reaps more benefits from *Plasmodium* if he or she is endowed with a portion of sickle cells. But eliminate *Plasmodium* from a place, as happened in much of the temperate world by the early twentieth century, and sickle cells are no longer adaptive but harmful, since they carry less oxygen. Sickle-cell diseases persist today across large swathes of India, the Arabian Peninsula, North Africa, Greece and Sardinia, and wherever else malaria was once pervasive.

We are quick to blame such diseases on humanity's long association with malaria parasites, but in places and times of widespread malaria, *Plasmodium* was for the most part just trying to do its sickle-cell carriers a favour.[18]

Centuries, indeed millennia, of exposure to malaria's parasites thus means that people from malarial and ex-malarial areas around the world still live with the memory of *Plasmodium* in their DNA, manifested as sickle cells, as well as various other genetic adaptations such as G6PD deficiency (or favism), scattered thalassemias and several other red blood cell traits. In 1961, a decade after Sardinia saw its last case of indigenous malaria, 14 per cent of islanders still carried the gene for thalassemia, while 30 per cent of them carried the gene for favism. These genetic conditions are ways that earlier Sardinians sought to accommodate and adjust to their mosquito-transmitted protozoan – thereby allowing themselves to take better advantage of the benefits that this parasite could offer them, such as fending off pirates and other marauders not endowed with these genes. One can also realize why many Americans of African heritage were, and are, less susceptible to malaria than Americans of northern European heritage. Although the sixteenth-century Spanish priest Bartolomé de las Casas was convinced that the first Africans brought to the Caribbean were protected from common diseases by their 'thicker skin' or perhaps their 'offensive odours', subsequent theorists credited African American resilience to their better acclimatization, that is, their better adaptation to hot and humid climates. Likewise, southern U.S. plantation owners had long observed that their slaves were much less susceptible to common fevers, attributing such differences to how the African tropics had crafted human bodies to withstand oppressive heat and high humidity. As Mart Stewart explains, not only was it assumed that the 'Negro' could work harder and longer under stifling conditions compared with other races, but that his or her body had been acclimatized to fend off agues, marsh fevers, flus and other maladies. Katherine Johnson qualifies that recently arriving enslaved people sometimes

The 9th U.S. Volunteer Infantry, a Black 'Immune'
regiment deployed to Cuba in 1898.

had to undergo a period of 'seasoning' as they adjusted to a different *Plasmodium* species from the ones they had been habitually exposed to in Africa. Likewise on the California frontier of the nineteenth century, explains Linda Nash, differential racial susceptibility to disease led bystanders to assert that malaria 'preyed' disproportionately on whites. Yet the level of resistance to malaria in Black Americans by this time was determined by the recency and potency of their acquired immunities coupled with any inherited adaptations to the disease. Many of them did, in fact, became quite ill with the fevers: generations and indeed centuries might have passed since their forefathers had been parasitized by plasmodia.[19]

The disadvantages of *not* being parasitized (or not being parasitized recently) were brutally demonstrated at the turn of the twentieth century in the struggle over Cuba during the Spanish–American War.

In this conflict, African American soldiers became the subject of a special recruitment campaign because of their supposed resistance to tropical diseases. Recruiting for what were popularly known as the 'Immune' regiments, the u.s. Congress issued a call for 10,000 men 'who, owing to their origin, the places of their residence, and other circumstances affecting their physical characteristics, possess immunity, or are likely to be exempt from diseases incident to tropical climates'. In the key Battle of San Juan Hill of 1898, a regiment of Black Immunes was sent to the front lines to rescue white soldiers debilitated there by yellow fever and malaria – but with the surprising result that many of the so-called Immunes soon found themselves suffering from the same diseases. As historian Marvin Fletcher recounts, 'Not a soldier could report for duty, few were able to prepare meals, and many were quite sick.' Generals had to confront the fact that immunities might very well wane in people who lacked continuous exposure to a parasite. If the aim was to reap the full benefits of hosting plasmodia soldiers with more recent exposure to the dreaded diseases would probably have made better recruits in this war. Freshly arriving immigrants from Greece or Italy, for example – many of whom had much whiter skin than the Immune regiments – would probably have made more resistant cannon fodder in these Caribbean battles since their Mediterranean origins were at that time still endemic with malaria. In their war strategies, it seems that u.s. Congressmen could have been less racially biased in recruiting soldiers than they might have guessed.[20]

The Benefits of Mutualistic Parasites

Beyond the benefits we reap from our pathogenic parasites are those we accrue from more directly mutualistic ones. Although a 'mutualistic parasite' may seem an oxymoron, since parasites are by definition taking and not giving (or hurting and not helping) and would seem unable to offer their hosts any advantages, there are in fact many parasites that provide their hosts with immediate and direct benefits that do not depend on promoting their host's resistance to disease. The challenge here may be one of terminology, since a creature initially labelled a parasite may with further observation be deemed a mutualist. Taking Goethe's dictum to heart that 'just when ideas fail, a word comes in to save the situation,' one can adopt the term 'mutualistic parasite' to clarify that a good many parasites provide their hosts with significant and immediate benefits.[21]

An important category of such parasites includes those creatures that can displace or neutralize more virulent parasites, for example, or else mitigate their ill effects. So while pathogenic parasites can help us by infesting and debilitating our enemies and competitors if we ourselves can resist them, mutualistic parasites earn their keep by providing us with greater defences against illnesses. It turns out that there is a variety of benign and benevolent parasites that mask or supplant our pathogenic parasites and pathogens – and we would be much sicker if we did not carry them. An excellent example of these mitigator parasites may be *Plasmodium vivax*, which, by causing milder forms of malaria, can in some cases serve as a prophylactic against contracting malaria's deadlier forms, especially those caused by *P. falciparum*. Once established in the human body, the less dangerous *P. vivax* apparently occupies niches that would be occupied by more virulent *P. falciparum*, even when both species of *Plasmodium* infect a person simultaneously. A person who accommodates *P. vivax* commonly suffers milder fevers and headaches, and is less likely to experience severe tertian (or 'pernicious') symptoms characteristic of *P. falciparum* infections that can lead to cerebral blood clots and death. A pair of malariologists go so far as to claim that '*P. vivax* somewhat "vaccinates" against the clinical severity of *P. falciparum*.' It has also been shown that around the world, more asymptomatic malarial individuals carry *P. vivax* than any other *Plasmodium* species, perhaps owing to the fact that *P. vivax* can tolerate larger ambient temperature ranges than *P. falciparum*. Moreover, molecular evidence reveals that *Homo sapiens* has been living with *P. vivax* for much longer than with *P. falciparum* – for some 200,000 years instead of just 10,000 years, by some estimates – lending credence to the view that this extra time of parasite–host cohabitation has allowed us humans to build stronger alliances with our more familiar *Plasmodium* parasite.[22]

The advantages of hosting the less virulent of two similar parasites becomes apparent if one carries *P. vivax* into a malaria-infested land, with these mitigator parasites being one of the ways that humans can enjoy greater resistance to a disease. Since the southern United States continued to witness malaria outbreaks into the nineteenth century, some of the descendants of the first African Americans with plasmodia in their veins were able to maintain malaria resistance for several generations. If one of these southerners then migrated to gold-rush-era California, where malaria topped the list of health problems, *P. vivax* could provide a smoother arrival for its host if exposed to this malaria, whether due to *P. vivax* or *P. falciparum*. During this time in the 1850s and '60s, malaria mortality rates for central California's indigenous peoples are estimated to have approached 50 per cent, with these rates also running high for the state's recently arriving European Americans. But malaria mortality rates for recently arriving African Americans were much lower. These settlers' ongoing malaria resistances meant that they could spend more time dealing with the other hardships of the frontier, and less time convalescing from this feverish disease. In fact, for your average Forty-Niner arriving in the gold fields, it was probably much less useful to be carrying an extra sluice pan or bag of flour in their backpack than a strain of *P. vivax* in their bloodstream.[23]

A rather curious and surprisingly successful medical treatment in the early twentieth century involved taking advantage of some of *Plasmodium*'s cruel symptoms. The savage effects of syphilis, which often followed sexual contact, would typically start with skin lesions, then progress over the months and years to infect the nervous system, eventually crippling both body and mind – sometimes severely. Perhaps the only really serious pathogen exported from the New World to the Old, syphilis bacteria spread across Europe to infect, according to some estimates, some 5 to 20 per cent of all inhabitants by the 1920s. Because of its method of transmission and progressively gruesome stages, it received more attention than tuberculosis, the day's leading killer. Medical experts feverishly sought any relief from syphilis or possible cures. In the days before modern antibiotics, the most promising anti-syphilitic treatment would come to depend on the effects of one microbe countering those of another. This new treatment was called malaria therapy.[24]

A counterintuitive remedy, the project of fighting syphilis with malaria was discovered and refined by renowned Viennese psychiatrist Julius Wagner-Jauregg. Before developing malaria therapy between 1917 and 1921, Wagner-Jauregg had focused his attention on 'pyrotherapy', or the induction of high fever in infirm patients, which in some cases was found to mitigate their symptoms. Various forms of fever therapy had been practised since ancient times. Wagner-Jauregg had observed the amelioration of certain psychoses under naturally contracted febrile illnesses, leading him to try to induce fevers in patients by artificially exposing them, first to strains of streptococci, later to tuberculin bacilli, and eventually to malaria parasites. A week or two after the patient was exposed to one of these fever-producing infectious agents, Wagner-Jauregg then administered an antidote to bring the patient out of his or her febrile stupor, sometimes witnessing significant improvements in the patient's psychotic condition. Through dubiously ethical trial-and-error methods, he found that malaria's high fevers gave the best results, with quinine then being an efficient remedy for ridding the patient of this disease. Malaria therapy provided amazing relief to many syphilitic patients for whom there had been very bleak prognoses. His records showed that, despite occasional complications and accidental deaths from induced malaria, the majority of his patients enjoyed partial or full recovery from their psychoses, so much so that many of them were discharged from psychiatric institutions to return to normal life. The inoculant of choice became *P. vivax*, since it produced a less dangerous form of malaria. Wagner-Jauregg won international recognition for his therapy and a Nobel Prize in medicine in 1927.[25]

In our own day, the growing field of 'evolutionary medicine' might explain the mechanism of malaria therapy as stemming from raising the body's internal temperature to become more damaging to the pathogen than to the human host. The painful, debilitating, disfiguring and neuro-degenerating symptoms of advanced syphilis, which follow long-term infection with *Treponema pallidum* bacteria, can be mitigated by hosting other bodily inhabitants, be they bacteria, such as *staphylococcus*, *Mycobacterium tuberculosis* or a protozoa such as *Plasmodium*. Decades after Wagner-Jauregg's discovery, researchers are still unclear about the precise mechanisms responsible for the successes of this therapy; they are certain only that malaria therapy can produce beneficial results, despite its potentially dangerous complications. As Wagner-Jauregg explained in his Nobel acceptance speech, 'The overwhelming majority of writers agree that with this method remissions can be obtained which are on a scale far exceeding those attained by any other method.' Data from 1941 showed malaria therapy remission rates as high as 50 per cent for patients suffering from syphilitic disorders. A San Francisco physician at the time added that malaria therapy, though contraindicated for delicate patients, proved useful for treating not only syphilis but other troubling conditions such as gonorrheal arthritis, epididymitis and certain urethral and prostatic infections. Here were modern physicians fighting one disease with another, demonstrating that a parasitic infection can help a person fend off other more troublesome infections.[26]

Although many syphilitic individuals benefited from malaria parasites after Wagner-Jauregg's discoveries, it should be noted that his treatment did not work very well on certain individuals, particularly those with African heritage – as might be suspected when one accounts for the effects of acquired and innate immunities. For many such individuals, it was reported that malaria would not 'take' (that is, would not be contracted) following malarial inoculation, which typically consisted of receiving a small injection of a malarial person's blood. Even after repeated inoculations or else confinement to a room containing both mosquitoes and malarial individuals, many of these syphilis patients could not reap the benefits of malaria therapy for the simple reason that they never became very malarial. Even many of those who were infected with more dangerous, higher-fever-producing *P. falciparum* were still unable to manifest the requisite fevers, as was brought to light in Alabama's infamous Tuskegee Syphilis Study, which was initiated in 1932 and lasted several decades. Despite this study's deplorable and racist methodologies, which subjected African Americans to a battery of questionable experimental treatments without their full understanding of the risks they faced, it was becoming clear just how malarial resistant certain individuals could be if they had been previously exposed to *Plasmodium*. The consequences of having been earlier exposed to this parasite, or of having one's near ancestors repeatedly exposed to it, could prevent a person from fending off more dangerous bodily inhabitants. Thus, in the case of malaria therapy, being parasitized could entail costs as well as benefits: when seeking syphilitic relief through Wagner-Jauregg's techniques, previous exposure to malaria was disadvantageous. A different fever-inducing treatment was needed.[27]

As biologists uncover the molecular pathways by which malaria therapy functions, one can appreciate that elevated body temperatures produce altered enzymatic activities that can hinder or eliminate infectious agents. One can also appreciate that any parasite, whether seriously damaging to its host or not, has little interest in ultimately killing that host, but a great deal of interest in killing other competing parasites that utilize the host's same resources of habitat or nutriment. The parasite that can survive the host's highest fever temperatures will win out in life's race. Little wonder that for most people across most of history, bouts of fever, usually benign but occasionally dangerous, have been normal parts of life, explains Christopher Hamlin, author of *A Short History of Fever*. As an eighteenth-century Dutch physician declared, 'No person can live without fever.'[28]

Interestingly, malaria therapy has more recently been investigated as a treatment for those suffering from other infectious diseases, including Chagas disease, leishmaniasis, poliomyelitis, AIDS and Ebola. At least one researcher during the COVID pandemic hypothesized that high body temperatures might combat SARS-CoV-2, the virus responsible for the disease. In all of these cases, researchers were speculating that it is better to suffer a mild infection from malaria than undergo a drawn-out battle with a more deadly microbe: they hoped that one bodily inhabitant would counteract the negative effects of another. It therefore seems clear that *Plasmodium* can be friend and not just foe. We humans are ecosystems, serving as a habitat and battleground for micro-organisms that are living, reproducing, competing and cooperating in our veins or bowels, and some of these creatures are providing us with vital services. As microbiologist and philosopher René Dubos explained it, 'living things constantly harbor a host of different kinds of microbes, some of which contribute to their well-being while others become the cause of disease under certain circumstances.'[29]

The Benefits of Curative Parasites

Malaria therapy is just one example of how a microbe can counteract the negative effects of another microbe, or even provide curative benefits. Humans purposely or accidentally ingest or absorb a range of organisms that can offer health advantages beyond inducing fever. As we have already found, cultivating a robust population of intestinal worms can alleviate or eliminate Crohn's disease, a debilitating auto-immune condition. A physician may prescribe capsules of 'probiotics' or even faecal transplants to a patient who has undergone repeated treatments of antibiotics for helping replenish that patient's intestinal biodiversity. There is also evidence for the advantages of so-called cross-immunities whereby exposure to one pathogen confers partial or full resistance to another. Thus various studies conclude that a person is less likely to catch the common cold if he or she is already battling the flu. In developing the first consistently effective vaccine two centuries ago, Edward Jenner built on the practice of variolation, by dripping or flicking liquid from cowpox pustules onto a person's skin lesions to promote a minor cowpox infection so as to provide protection against future smallpox exposures; Jenner thereby demonstrated how a relatively benign virus can counter a very malignant one. In fact, one of the main methods of immunization relies on exposing a person to a less harmful pathogen that is similar (and often attenuated) but not identical to the target pathogen. Many of these cross-immunities are unplanned, whereby a person enjoys resistance to more dangerous diseases because of former, accidental exposures to a pathogen. To highlight an example in medieval Europe of this kind of fortuitous exposure, many *Salmonella* carriers apparently enjoyed resistance to plague. Similarly, those who contracted leprosy may have also enjoyed protection against plague, says Richard Hoffman, since so many lepers were able to skirt the Black Death – and did so to such an extent that 'some lepers were persecuted and killed on suspicion of propagating an illness which they seemed to defy.'[30]

There are multiple examples of human beings reaping the curative benefits of their bodily creatures, even when such creatures are pathogenic. One of the central mysteries of the deadly Spanish flu epidemic of 1918–20 is that young adults in their twenties and thirties were especially susceptible to the disease, even if the elderly are typically much more vulnerable to such viruses. Many parts of the world lost 4 to 5 per cent of their citizenry to this flu. But instead of pointing blame at the Spanish flu's unusual pathogenesis, historical epidemiologists have uncovered evidence for the existence of similar though more mild strains of this virus that swept the world two generations earlier, thereby conferring immunity protections on those born before 1880. Since people born after this date were not exposed to these strains, they missed out on acquiring the crucial immunities that may well have defended them against the more lethal flu varieties that arrived in 1918. Here again, it was better to be sick when young than deathly ill in early adulthood.[31]

In our own day, one can point to a study showing that Senegalese children were more likely to survive a battery of early-age illnesses if they had first contracted and survived measles. There are also indications that a person's ordinary and healthy *Escherichia coli* bacteria can provide a degree of protection against the discomforts brought on by malaria parasites. And to flip our view of malaria parasites from pathogenic to curative, there is evidence that plasmodia may themselves provide protection against a range of auto-immune diseases. These observations should make us suspect that any general programme of bodily parasite eradication would probably also serve to eradicate a number of known and unknown benefits with which these parasites are providing us.[32]

In still more roundabout ways, our microbial symbionts can provide some of us with the advantages of 'herd immunity', which protects a minority of individuals if a majority of hosts has already been infected. There is the cautionary tale of those who contracted polio as young adults because improvements in sanitation meant that they were not exposed as infants to the poliovirus that could have been neutralized with maternal antibodies. Much like Spanish flu victims, those polio victims who waited to confront a dangerous virus later in life would eventually suffer even more severe consequences. Interestingly during the SARS-CoV-2 pandemic, there were even recommendations to re-inoculate people with polio vaccines: with coronavirus infections spiking across the world in the summer of 2020, virologists acting on evidence announced that surplus polio vaccines, safe and plentiful since polio no longer circulated, might provide significant protection against COVID-19 and its complications. Since poliovirus and coronavirus share a similar genetic structure, it was felt that the polio vaccine would stimulate the secretion of interferons useful for combatting both viruses. As a group of virologists reasoned, 'emergency immunization with live attenuated [polio] vaccines could be used for protection against other unrelated emerging pathogens.' To test the claim, a regiment of frontline coronavirus healthcare workers were inoculated with polio vaccine; they went on to register moderately favourable results in fending off the virus.[33]

All these examples show how such issues as timing of parasite exposure, size and distribution of host population, and degree of genetic relatedness between parasites can all be crucial factors for determining how far one can benefit from one's parasites. Such spatial and temporal differences, and the resulting collective or indirect consequences, mean that even very dangerous parasites are rarely simple spectres of death. One must remember that it is in the parasite's best interest to make sure that its host ultimately survives, thrives and goes forth to multiply so as to create better opportunities for itself. A benign host–parasite union is often advantageous to both members, regardless of what many practising parasitologists feel about the parasites they hope to eradicate. As Pierre Joseph van Beneden taught long ago, a parasite interacts with its host 'but prudently, so as not to endanger his life'. Although a parasite's prudence may be difficult to describe or quantify, one may well locate it in the various benefits that it can bring to the host.[34]

The Benefits of Mind-Altering Parasites

To turn to some of the more unusual and often spectacular benefits of being parasitized, one can pass from a parasite's immediate physiological effects to consider how it may alter a host's behaviour or judgement. So-called mind-altering parasites have of late become popular objects of study, and here again, there is a range of ways by which they may be providing advantages to their host. A famous example of one mind-alterer is *Toxoplasma gondii*, a single-celled protozoan that is about as large as a malaria parasite, though even more widespread. The world is filled with these creatures, since they are able to infect most warm-blooded animals, including people and their domestic pets, especially cats, making them common in households and wherever else our intimate animals roam. These parasites act by infecting nervous tissue and altering neurochemistry for modifying the actions and movements of their hosts, sometimes producing cysts that get lodged in the brain. In several well-documented studies, a rat or mouse that becomes infected with *T. gondii* often develops a mysterious attraction to cat odours, making these rodents more likely to approach cats, and so become more susceptible to cat predation. It turns out that *T. gondii*'s life cycle requires that it spends time within a cat's digestive tract to fully mature before being returned to the soil via the cat's faeces to be picked up by another rat and then cat. Some researchers claim that this parasite is 'controlling' the mind of its rat host through a chemical process that makes it perceive cat odours as sexual stimulants, thereby convincing rats to steer straight into cats. The question then becomes if, or how, *T. gondii*'s zombie-control of the rat could possibly be providing benefits to this host – since the parasite's odour trickery sends rats to their doom.[35]

As with most things parasitological, one must search deeper to unravel how a host may benefit from its parasites. In one explanation to the paradox of the parasite that produces suicidal rats, it seems that female rats prefer to mate with *T. gondii*-infected male rats, which it can better detect through smell, thereby increasing the infected male rat's fitness despite it becoming easier prey for a cat: here, *T. gondii*'s manipulation of the rat's sense of smell is really for the rat's own benefit. Yet a more intriguing explanation of the suicidal rat focuses not on rat but on cat, for it seems that *T. gondii* is more interested in pleasing felines than rodents. By sending a rat into the jaws of a cat, *T. gondii* is serving up the rat (the intermediate host) to the cat (its definitive host) – with *T. gondii* requiring vital developmental stages in the rat as well as the cat. In what biologists term 'extended phenotype', the parasite enlarges its world by making life easier for a cat, which more readily catches a zombie rat to then pass itself on in cat faeces to attract another scavenging rat – which is then eaten by another cat. In this cycle, the parasite's main beneficiary is the cat, with the rat being sacrificed for the cat that makes crucial habitat for the parasite. Or it may well be that the parasite is benefiting both rat and cat by helping the former find a mate while assisting the latter to keep its stomach full.[36]

Such biological complexities are relevant not only to biologists, since various multiple-host-parasite relationships extend to the human world as well. It turns out that the malaria parasite may also be employing this sacrificial strategy of seeking to please the mosquito (its definitive host) even more than the human being (its intermediate host). Remembering that malaria is a *ménage à trois* – or three-species affair – means that from the perspective of the malaria parasite, humans act as so much fodder for ultimately satisfying the mosquito: after all, the mosquito is fundamental for transporting this parasite to the next human. *Plasmodium* requires both humans and mosquitoes to complete its life cycle, but it is only the mosquito that can provide the parasite with wings for transport. There is intriguing evidence showing that *Plasmodium* stimulates humans to produce distinct odours in the blood that emanate through the human skin to attract mosquitoes, since mosquitoes are observed to preferentially feed on malarial individuals over non-malarial individuals. So even if there are various direct and indirect ways already mentioned for how *Plasmodium* can benefit people – as through pathogenic, mutualistic and curative means – one realizes that *Plasmodium* is also deeply interested in pleasing the mosquito. If at one level, humans are merely providing foodstuff and habitat for the parasite, then the main beneficiary of being parasitized would be the mosquito – its second, mobile host. Mosquitoes will be better fed and lay more eggs because they can more easily locate their next blood meal through the help of the malaria parasite. In the end, *Plasmodium* would seem to be giving back to mosquitoes as well as to humans – repaying both for the advantages it gains, and ultimately calling into question just how parasitical this parasite is.[37]

In the case of *T. gondii*, we can add that this microbe is also infecting people, not just rats and cats – although, not surprisingly, especially people living with rats and cats. Currently, it is estimated that some 10 to 20 per cent of those residing in Western countries are infected by *T. gondii*, reaching twice this number in tropical countries, but with cat-loving regions, say, of rural France showing evidence that almost half of all residents share their bodies with this parasite. Although these infections are often asymptomatic, some doctors blame a range of serious illnesses on this protozoan, from encephalitis and ocular diseases to epilepsy, schizophrenia and bipolar disorder. With *T. gondii*'s primary site of infection being brain and nervous tissue, the associated pathologies reflect the altered functions of these organs. Data show that individuals suffering from these mental illnesses generally do have much higher levels of *T. gondii* infection – even if one researcher qualifies that 'there is not currently sufficient evidence to indicate these associations are causal.' Other studies point out that *T. gondii*-infected humans (like rats) may become easier prey for feline predators, since there is evidence showing that leopards more easily prey upon *T. gondii*-infected chimpanzees (a close primate cousin of humans) that are attracted to leopard odours. Still other researchers simply insist that *T. gondii* is a dangerous pathogen, causing disease and suffering.[38]

Yet the fact that *T. gondii* can live inside so many humans asymptomatically means that we do not yet know the whole story of this parasite–host union. It may not be surprising, then, to find out that this parasite may well be providing a range of side-benefits to its human host despite its links to several pathologies. For one, there are strong indications that these protozoans provide their hosts with resistance to such troubling auto-immune problems as multiple sclerosis, thereby lending further credence to the hygiene hypothesis. Add to this possible benefit the observation that latent *T. gondii* infection induces greater dopamine secretion in its human host, which may give that person 'paradoxical improvements of cognitive control processes'. Elsewhere, a large sample of toxoplasmodic individuals was found to be more adept at shifting their attention from one subject to the next, so as to improve their 'action control which contradicts the commonly held view that parasitic manipulation is always to the detriment of the host'. More controversial studies suggest that toxoplasmodic men are greater risk takers, a behaviour that may provide them with advantages in life's competitive race – or that toxoplasmodic women are more than twice as likely to give birth to sons than daughters, a possible reproductive advantage since men can theoretically produce more progeny than women. Still other investigators present evidence to suggest that *T. gondii*-parasitaemic individuals show higher aggression (in women) and higher suicide risk (in men). Whether we are willing to accept all of these conclusions with a straight face may be beside the point, since it is easy to suspect that spurious correlations are at work here – or else that these gender-weighted conclusions are better explained by gender-skewed compositions of the study teams.[39]

Leaving aside questions of scientific objectivity, one can nonetheless witness an emerging school of thought that maintains *T. gondii* is clearly pathogenic – composed mostly of medical practitioners – which contrasts to a diverging school that presumes that these parasites produce net-favourable effects for their hosts – composed mostly of ecologists and psychologists. We are then left to conclude that, in our search for understanding long-term and/or indirect human consequences of being infected with *T. gondii*, more research is needed. In the meantime, we can continue reading studies, and find several that indicate that many of our bodily microbes really do change the way we act and think, such that

> the gut microbiome can affect behavior through interactions with the host neuroendocrine system. Some of those behaviors include stress-related behavior, social behavior, sexual behavior, cognition and addiction, all of which are modulated by neuroendocrine pathways.

And so while one cannot yet assume that most humans, under most circumstances, are benefiting from the ubiquitous *T. gondii*, a great deal of evidence from other parasites infecting other hosts suggests that we should not eliminate this possibility.[40]

At this point, one can call attention to Michael Pollan's 'botany of desire', whereby garden-variety plants, such as potatoes, tulips, apples and even marijuana, are manipulating us to plant them because we are attracted to their tastes or beauty or intoxicating effects. People plant delicious potatoes or beautiful tulips not because we want to plant them, argues Pollan, but because these plants are convincing us with their foodstuffs and their flowers and their effects that we should go out and plant them. Plants are in control. Gardens flourish not because of the motives of gardeners, but because of the manipulative powers of plants. It may then follow that in the world of microbes, zombie-controlling parasites are also using this strategy in their project of manipulating their hosts. If a plant can persuade us to sow and water it, think what a savvy parasite inside our bowels could convince us to do! As seen earlier, parasites can stimulate one host to exude odours to attract another host, as when *T. gondii* dangles rats in front of cats – in the same way that *Plasmodium* may be dangling humans in front of mosquitoes, thereby allowing themselves to thrive and reproduce in complicated multi-host life cycles. Much like our garden plants, our parasites are providing their definitive hosts with the benefits of easy food for better propagating themselves. The other result is that, while the gardener reaps advantages from her plants, the mosquito reaps advantages from her *Plasmodium* for ultimately benefiting both. *Co-culture* is a recent term coined by Cédric Sueur and Michael Huffman to denote a union of organisms whereby 'the behaviours of one species actively shape the cultural evolution of another and vice versa,' be it a bottlenose dolphin (*Tursiops aduncus*) herding fish into human fishing nets, or honeyguide birds (*Indicator indicator*) leading people to honey-filled hives, with both humans and non-humans sharing in the harvest. Parasite and host would seem an especially pervasive

example of co-culture, with their tight relationships seeming to benefit all involved in truly mutualistic unions. Only by assuming that a
Plasmodium takes more than it gives do we label it a parasite. Only by
uncovering the full nature of a parasite, its behaviour, its life cycle and
its effects, do we realize that it is hardly parasitic.[41]

And so why is it that we humans do not manage to exit the coercive
relationship created by the *Plasmodium* and propagated by mosquitoes,
which can give us raging fevers? When we walk along a creek or through
a forest while being pestered by a trail of mosquitoes, some of which
may carry the debilitating plasmodia it may seem that we can never
escape our three-member liaison, even if we are armed with powerful
insecticides and strong antimalarial medicines. But the answer to the
tight bond of our relationship lies in recalling the many advantages that
all three members can reap in this union. Although we may not care
much about the advantages we give to a *Plasmodium* or to the mosquitoes that carry them, we need to remember the benefits they give us:
there are indirect benefits that we gain from pathogenic parasites; there
are direct benefits that we reap from mutualistic parasites; and there
are still more benefits from our curative parasites – and possibly from
mind-altering parasites. The *Plasmodium* that sprouts wings by living
inside a mosquito to then be injected into the next unsuspecting person
has provided ourselves and our ancestors – as well as the mosquito –
with a variety of advantages, although we are loath to acknowledge
them. Humans teeming with parasites have conquered their enemies,
displaced more perilous microbes, rid themselves of diseases, aided
bodily functions and just maybe created more favourable mental states.
Our donation of five-millionths of a litre of blood during a typical
mosquito bloodmeal, which serves to nourish 10,000 thriving plasmodia – that may go on to produce seasonal fevers – would seem the
least we can do to repay our wriggling little partners.[42]

Parasites in Daily Affairs

Richard Hoeppli has searched extensively through ancient Chinese medical records to uncover a rich folk tradition involving parasites in human affairs. Flatworms, roundworms and lice surface frequently in these early accounts, with shamans and healers blaming various human maladies and discomforts on these bodily creatures. But just as often, these early health authorities praise our various intimate inhabitants for their role in promoting recuperation, good health and good spirits. One of Hoeppli's archival nuggets mentions, for example, that 'If one has a great number of short worms in the abdomen, one often dreams of crowds of people, while [one who harbours] many long worms frequently dreams of fighting with other people.' In determining whether these worms could bring harm or benefit, their numbers and densities were usually as important as the timing of their infection. Western folklore is found to serve up similar parasitic wisdom over the centuries, with various bodily creatures, such as blood-sucking leeches and even maggots, being considered important fixtures in the physician's toolbox. A military surgeon in the u.s. Civil War, when faced with a soldier's festering battle sores, declared that maggots could in a single day 'clean a wound much better than any agents we had at our command'. Leeches, perhaps more predatory than parasitic, were employed initially for withdrawing liquid for rebalancing the body's humours; they were eventually used for sucking out blood thought to be laden with noxious agents. One source claims that during each year of the 1830s, 5 to 6 million leeches were utilized in Paris alone, revealing just how many patients must have looked on appreciatively as *Hirudo medicinalis* was draped over their skin to began swelling with blood. George Washington during his last days would probably have preferred bloodletting through the more painless alternative of a hungry leech attached to the skin than his actual fate of a sharp knife drawn across strategic veins. It is easy to forget that our body's creepy crawlies have often been viewed as allies and companions. Still today, if we have a special craving for bread or cheese or ice cream, some trace the root cause to a worm-lined GI tract.[43]

Such anecdotes, sometimes corroborated by archival evidence, reveal that the average shepherd or blacksmith or milkmaid simply learned to live with their wriggling creatures, considering them as nuisances if burrowing under their fingernails or between their toes, but perhaps also offering mild celebrations of them, especially if they were considered to have curative properties. Hoeppli relays one Chinese proverb that 'man should have at least three worms in order to remain in good health.' Across Europe, intestinal worms were considered useful for stimulating digestion (as through peristalsis) or for consuming excess mucus that might otherwise 'putrefy', particularly in children. Van Beneden pointed out that the Abyssinians of east Africa 'do not consider themselves in good health, except when they nourish one or many tape-worms'. Other health experts insisted that a rash of scabies (stemming from skin mites) gives a measure of protection against fevers and shortness of breath as well as mental diseases. Even today, a Chinese physician recounts how doctors in her country have long prescribed mixtures of faecal bacteria and urine, called Yellow Dragon Soup. 'It was used to treat diarrhea,' she explains. These and other snippets about the virtues of parasites sound eerily familiar to the arguments of current medical researchers such as Martin Blaser, who claims that our own day's missing microbes, stemming especially from the overuse of antibiotics and other parasitical scrubs along with hyper-hygienic lifestyles, are crucial reasons why so many of us suffer precisely from such conditions as asthma and stress, even diabetes and obesity.[44]

Even in the case of lice, the little arthropods found attached to hair follicles in our scalp and groin, absence seemed to signal imbalance and sickness. Microbiologist Hans Zinsser extolled the value of these creatures, noting that 'long into the eighteenth century, lice were regarded as necessities . . . as wise a man as Linnaeus suggested that children were protected by their lice from a number of diseases.' An

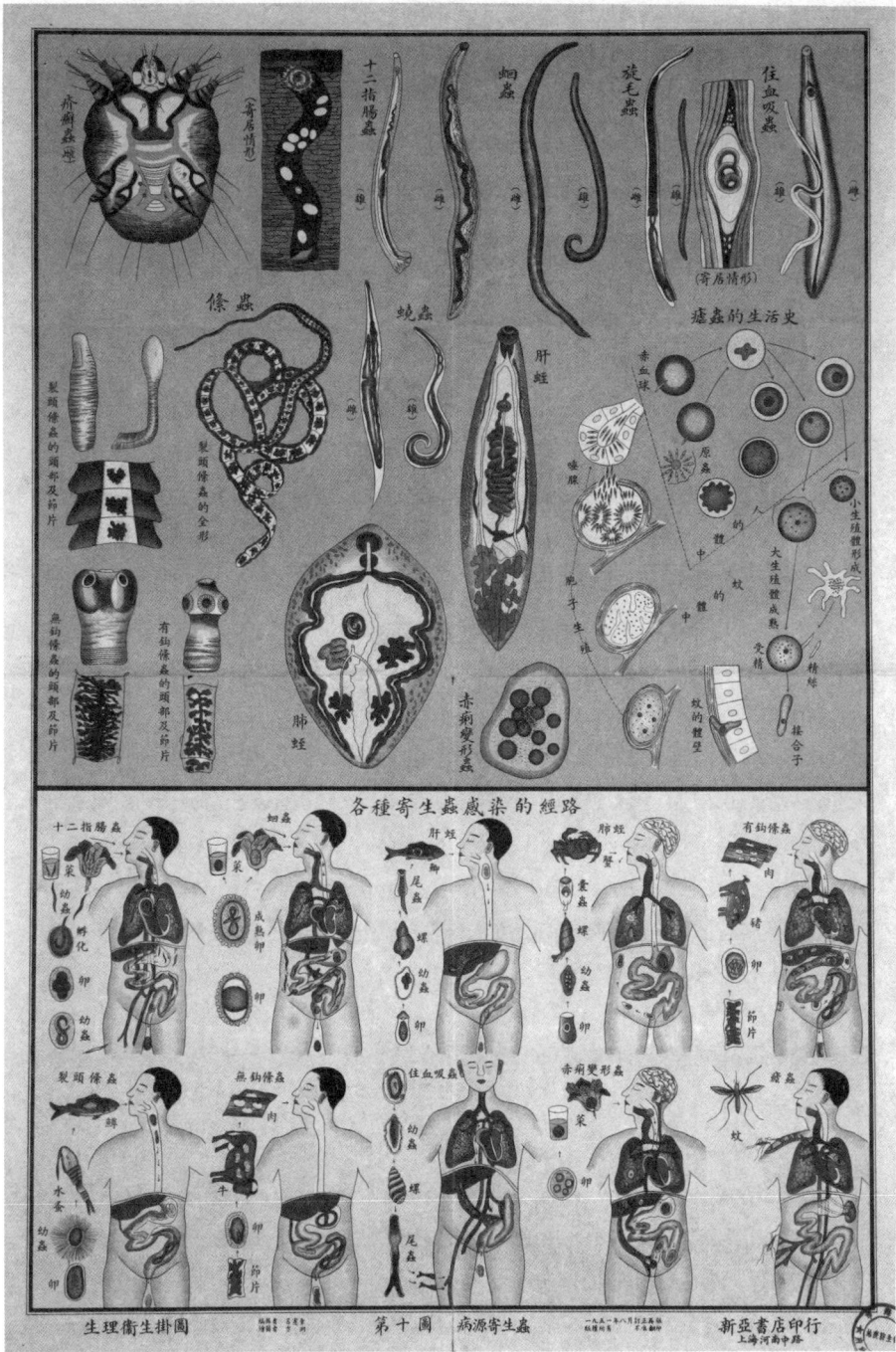

Chart of human parasites and their modes of transmission,
by Xianzhang Lü and Dong Fang, 1951.

abundance of lice in the pubic area was taken to be a sign of good health, even virility. Contrariwise, congenitally infirm and elderly people might show an unusual scarcity of bodily parasites, especially lice. In fact it was well known that lice would at some point abandon the seriously infirm, with the little critters fanning out from a moribund patient to seek more vigorous bodies to inhabit. Modern-day immunologists focusing on lice note that these creatures 'may be a strong source of immunoregulatory stimuli that is absent in many modern human populations'.[45]

Indeed, any veterinarian or helminthologist today realizes that to obtain live, healthy intestinal worms, one must collect them immediately after the death of the infected animal: should intestinal worms be left for very long inside a slaughtered cow or sheep, the worms will rapidly decompose through the actions of the host's digestive fluids. Such parasites are apparently quite content as long as the host is alive – and indeed it seems that a living host somehow protects its intestinal parasites from its own corrosive juices. Another observation at the laboratory reveals that mice sterilized of their intestinal creatures are found to live shorter lives than those that maintain a rich microbial diversity. Both historical and contemporary sources in laboratories and beyond suggest that we hosts are often much better off if we can maintain and even cultivate a rich variety of fellow travellers.[46]

Deterministic Parasites

In reviewing the many and varied benefits of being parasitized, there is certainly the risk that one can read too much into our nematodes and trematodes, mosquitoes and protozoa, *Escherichia coli* and *Yersinia pestis*. Dusty folk traditions may be giving these oozing and buzzing creatures far too much agency in human affairs, even allowing us to flirt with a kind of 'parasitic determinism' whereby the mere presence of a viable parasite guarantees this or that human behaviour or serves to clear up a particular health problem. The occasional tapeworm or random slurry of intestinal microbes that has taken up residence inside us may have rather unremarkable or prosaic functions compared to, say, those of the little snail worm (*Leucochloridium spp.*) that finds its way inside a snail's tentacle (*Succinea putris*) to begin pulsating so excitedly that it attracts an overflying sparrow (*Passer domesticus*), convincing the bird to gobble up the snail so that the worm develops further within the bird and is then passed back to the soil, reincarnated and ingested by another snail. The pulsating worm exerts powerful influence over the destinies of both snail and bird. But can we really make such dramatic claims for the parasites found in and on us?[47]

Some researchers insist that parasitic manipulation really is integral to the human condition. In the case of *Toxoplasma gondii*, one psychiatrist claims that this parasite 'might indeed be capable of transforming us into zombies of a sort with compromised free will'. Extremists assert that we are all as marionettes – essentially walking bundles of microorganisms – controlled by myriad strings attached to more than three hundred species of bodily worms, seventy species of protozoa and untold numbers of much smaller creatures living inside us and across our bodies. Marine ecologist and *T. gondii* determinist Kevin Lafferty declares that even if this ubiquitous cat-transmitted parasite can explain 'only' part of humanity's variation in 'two of the four cultural dimensions', his data nonetheless indicates that 'infected women are more intelligent, rule-conscious, dutiful, conscientious, conforming, moralistic, staid, rule-bound, warm, outgoing, attentive to others, kindly, easy going and participating' – with men manifesting their own battery of *T. gondii* counter-traits, most of them in their own contexts conferring advantages to their hosts. With so many positive effects of being infected by this cat-transmitted parasite, it may really be time that each of us adopt the nearest stray kitten, let it purr against our ankle and lick our cheek, even invite it into bed to curl up against our pillow so that we can assure ourselves of becoming toxoplasmic – if we aren't already infected.[48]

Even if it turns out that some of our bodily creatures do *not* produce many, or any, immediate tangible benefits, we cannot ignore the fact that even pathogenic parasites have offered advantages to us or our ancestors in very roundabout ways. There is the thesis of historian Walter Scheidel, for one, who believes that violence, warfare, famine and epidemics are the most effective ways for shaking up a society and redistributing its patterns of wealth. Through the logic of capital, he explains, and the institutions it spawns, society's riches soon become concentrated in the hands of the few, and it is only through periodic natural and social disruptions and catastrophes that these rigid class structures can be dissolved. Extensive international struggles such as the Second World War have been one such levelling event, he explains, but so have been the mighty epidemics such as the Spanish flu of the early twentieth century and the Black Death of medieval Europe. After *Yersinia pestis* mowed down a third of Europeans of all social classes in various waves during the 1300s, large estates were left without heirs, land rents and interest rates fell, and labour scarcity raised wages for the working classes. Here was a bodily inhabitant that could kill whole villages, carried in by fleas, themselves carried in by rats. Likely exported from China through central Asia via an early virgin soil epidemic, this bacterium went on to disrupt the very social fabric of Europe but with the result that it also 'flattened material inequality'. Similarly in the dramatic case of syphilis, explains William McNeill, poor and rich alike were infected, thereby preventing 'royal and aristocratic families [from giving] birth to healthy children', serving to accelerate 'social mobility, making more room at the top of society than there would otherwise have been'. Scheidel's and McNeill's reasoning thus points to a crucial silver lining of infectious disease, since by many measures, the marginalized classes were relatively better off a few generations after an epidemic than before it struck.[49]

So even as there is important evidence showing that the plague and syphilis killed the poor and sick disproportionately during the first generations, it is also likely that *Y. pestis*'s and *T. pallidium*'s longer-term consequences were not exclusively detrimental to the less fortunate classes. Disease agents spread across many generations tended to share their spectres of death more evenly, reaching over the barricaded walls of the privileged, sometimes with little distinction as to race, class and gender. Deadly epidemics must therefore be evaluated on individual as well as on collective scales, over small as well as large expanses of space and time. For the majority of citizenry, when judged over a period of centuries instead of decades, the aftermath of pathogenic microbes can reveal other benefits of being parasitized. As Jonathan Kennedy concurs, 'infectious diseases have destroyed millions of lives and decimated whole civilizations, but the devastation has created opportunities for new societies and ideas to emerge and thrive.'[50]

Before turning to humanity's repeated efforts to eliminate our most hated parasites, and seeing how the quest for eradication may have created more problems than solutions, I reiterate the puzzling observation that the once malaria-stricken island of Sardinia produces some of the oldest humans on the planet. If rampant malaria was the island's most notorious trait during the last century, human longevity may be its most famous trait in our own. Malaria's *Plasmodium*, which has been travelling inside us for almost as long as we have been human, circulates in our bodies and feeds on the haemoglobin in our blood cells, where it can produce fever, convulsions, loss of blood – even brain damage, coma and worse. Yet alongside these desperate symptoms, one needs to ask whether being parasitized with this protozoan is necessarily and always harmful, which is the accepted medical view. To bring new scrutiny to this assumption, one can remember that most *Plasmodium*-infected individuals remain asymptomatic or experience only mild discomfort. One can also remember the unexpected silver linings of malaria's debilitating effects, such that malaria parasites can produce fevers that assail other dangerous microbes, or aid their hosts in migrating to new lands and possibly defend them from immediate or subsequent invaders – even promoting greater social and economic equality. It is therefore worth coming to grips with the ultimate effects inflicted by our parasites, on individuals and on groups, now and in the past. Even true parasites – those that habitually take more than they give – may be offering some compensation and cooperation with the host that feeds them, since so much of their own survival depends on the good health and robust longevity of their partners in life.

4

Dreams of Eradication

The total effect that parasites have on man's health and food
is incalculable and their control or eradication is essential
for his wellbeing.

F.E.G. COX, in *Parasites, Their World and Ours* (1982)

'Now, a drug called chloroquine – and some people would add to it *hydroxy* – hydroxychloroquine', announced U.S. President Donald Trump, holds 'tremendous promise' for controlling the COVID-19 outbreak. 'I think it could be a game changer and maybe not. And maybe not. But I think it could be, based on what I see, it could be a game changer. Very powerful.' Added Trump: 'it is known as a malaria drug, and it's been around for a long time and it's very powerful. But the nice part is, it's been around for a long time, so we know that if it – if things don't go as planned, it's not going to kill anybody.'

In the hours and days following the president's press conference of 19 March 2020, the ensuing rush to corner chloroquine pills ticked up its price, emptied pharmacy shelves and led to hoarding. The U.S. Food and Drug Administration, in anticipating this frenzy, had already sequestered some 29 million doses before realizing that those already taking this medication for its proven alleviation of lupus, arthritis and a few other conditions were finding themselves in short supply. The rush was also on to substantiate Trump's claims, with people from soothsayers to university scientists becoming increasingly sceptical of this wonder drug as their investigations proceeded. As one Harvard physician tweeted a couple of weeks after the big reveal, 'Patients with lupus, arthritis, other conditions are *already* on hydroxychloroquine. And we are diagnosing them with Covid-19 LEFT AND RIGHT.' Another critic added that the 'Chloroquine hype is derailing the search for coronavirus treatments.'[1]

Certainly one would need to think twice before swallowing a drug typically used against a eukaryotic *Plasmodium* in order to fight a virus – an entirely different life form many powers of ten smaller and relying on different biological strategies. At first blush, battling coronavirus with chloroquine (or with its milder form, hydroxychloroquine) would be akin to launching anti-tank missiles at a swarm of houseflies. One would expect a lot of collateral damage. In fact, shortly after Donald Trump followed up his announcement with a tweet that continued extolling the virtues of this malaria remedy, physician and researcher Dena Grayson warned in a counter-tweet that this drug had occasionally caused fatal heart rhythms. It was apparently *not* true that Trump's elixir was 'not going to kill anybody'.

In the battle against dangerous creatures living inside living creatures, a pharmacist's central challenge has always been to find a remedy strong enough to wound or eliminate a pathogen without wounding or eliminating its host. Like the farmer who hopes to kill the pest but spare the crop, a doctor aims to prescribe medications with highly targeted effects. But the negative side-effects of even our most common pharmaceuticals can never be avoided. Drugs harm more than pathogens, just as pesticides harm more than pests. In its use against malaria, hydroxychloroquine has shown remarkable successes, but it can also cause chest discomfort, blistering of skin, blurred vision, decreased urination, difficulty in breathing, dizziness, fever, headaches, muscle pain, loss of hearing and unusual bleeding, among many other complications. No wonder some malaria sufferers prefer harbouring the parasite over ingesting the antidote: hosting *Plasmodium* protozoa can prove to be the lesser evil than receiving yet another dose of a semi-toxic antimalarial. Indeed, a few of the president's unlucky followers discovered a few days after his enthusiastic endorsement that it would have been better to continue living with their coronavirus than to have died from the complications of the trumpeted remedy – since deaths stemming from hydroxychloroquine consumption did occur. It turns out that another of Trump's impromptu suggestions for treating the coronavirus, which was unveiled a month before the hydroxychloroquine declaration and involved (theoretically speaking) a quick rinsing of a person's lungs with bleach 'by injection inside, or almost a cleaning', could not have brought worse results than fatally overdosing on hydroxychloroquine.[2]

With this chapter, I contrast the effects of hosting parasites with those that often arise from efforts to control them. My aim is to emphasize that parasitaemia, as the condition of harbouring parasites, is often less disruptive and less dangerous – to our physiologies and our societies – than the project of setting out to destroy them. I highlight experiences from around the world before returning to Sardinia to illustrate the side-effects from an epic battle against the malaria parasite, which expanded to encompass a battle against the parasite's transmitter, the mosquito. Extermination methods in Sardinia harnessed expertise of medical practitioners, pharmacologists and public health departments before enlisting ecologists, entomologists, land managers, regional planners and chemical engineers. Exterminators of parasites and of transmitters of parasites utilized every tool available at the time, successfully beating back malaria, but also transforming biological systems and human systems – and not always for the better. When the dust finally cleared in Sardinia, with the medications and pesticides drifting away, health workers had eliminated nearly every malaria parasite on the island, but at significant costs to land and life. In Sardinia's epic quest for eradication, one wonders whether local inhabitants may have lost more than they gained.

Gin and Tonic, Please

The president's excitement over chloroquine did serve to remind researchers that rigorous studies had been carried out on this drug many years before the COVID-19 pandemic that, in some circles at least, led to hope in its ability to kill the deadly virus. The earlier SARS-CoV-1 outbreak lasting from 2002 to 2004 – which also caused Severe Acute Respiratory Syndrome (SARS) and also arose from a coronavirus, albeit one with a slightly different molecular structure than SARS-CoV-2 – had caught the attention of a few ambitious researchers. Appearing in print right after the first SARS outbreak dissipated, their 2005 study reported that 'chloroquine has strong antiviral effects on SARS-CoV infection of primate cells . . . suggesting both prophylactic and therapeutic advantage.' Coming at a time when threats from the earlier pathogen were already waning, the study was mostly ignored, motivating scant follow-up research. Advisors counselling Trump pointed to this early study as justification for their own enthusiasm for the malaria drug. But as the COVID-19 pandemic wore on, renewed investigations clarified that the 2005 study held much validity but probably little applicability: chloroquine's success against both forms of coronavirus was limited to *in vitro*, laboratory specimens composed of monkey cells, specifically African green monkey kidney cells. The drug showed little efficacy at countering human coronaviral infections, even if a few dissenters maintained that chloroquine could provide therapeutic advantages to COVID-19 sufferers. At the height of the controversy, the discovery of a bogus data set led to a *retraction* by at least one prestigious publication, which concluded chloroquine to be unsuitable for treating COVID-19.[3]

It turns out that chloroquine had been around for decades. The story goes that Hans Andersag, a German-speaking Italian working in a Munich laboratory, first synthesized this drug in 1934 as a possible remedy against malaria, but initial clinical trials showed it to be too toxic to be of any real use in humans – that is, until a second investigation a decade later countered that this drug could in fact be an effective antimalarial with manageable side-effects. In the ongoing search for more effective drugs against humanity's age-old scourge, a typical strategy was to synthesize chemicals similar in structure to quinine, which was the main proven success story against malaria, being an extract of cinchona bark. Europeans travelling to Peru in the seventeenth century had observed indigenous Americans countering their fevers by drinking water soaked in this bark, and so adapted this practice to alleviate their own problems with the disease. Only abundant cinchona groves were required to keep the supply of quinine coming.[4]

Since no drug is perfect, and relapses of malaria fever – or ague, as it was commonly called – often raged despite the patient taking repeated doses of quinine, the extracted powder was dissolved and tested at different concentrations or mixed with various substances in the hopes of multiplying its therapeutic benefits. In 1828 one Dr Harty of Dublin combined quinine with mercury, for example, and although two of his test patients (borrowed from the local prison) salivated profusely, he reported that their ague soon disappeared: 'With one the disease never returned, with the other it reappeared after an interval of fifteen or sixteen days.' In the coming decades, quinine water would be combined with other substances for perfecting a palatable malarial 'tonic'. It is generally accepted that British colonists in India were the first ones to combine their own bitter tonic water with various herbs, along with strong spirits, especially gin, to make their beverage more appealing at the end of a long day. The gin and tonic was therefore born as a daily dose against malaria, being the fruit of the ongoing search to ward off the pesky protozoa by an agreeable means. The next century's synthesis of chloroquine (and then hydroxychloroquine) can be considered another step in the long line of remedies aimed at defeating the *Plasmodium*. Meanwhile, at the other end of the world, traditional Chinese herbalists found qīnghāo (*Artemisia annua*) to be a reasonably useful remedy against malaria, although its active ingredient of artemisinin was not rediscovered and brought to the world's attention until the late twentieth century. Tu Youyou's methodical search through

century-old health records, which led to her identification of artemisinin, along with her subsequent Nobel Prize in Physiology or Medicine,
are other testaments to the benefits of studying our parasitic pasts.[5]

With tonic water flowing freely by the mid-nineteenth century,
malaria healers tried substituting their gin with whisky, wine or rum
– depending on their favourite inebriating beverages – while being
attentive to optimal doses. If the tonic was too weak, lacking sufficient
quinine, one risked missing out on its curative potential and contracting ague anew. Yet if one mixed in too much quinine powder, the gin
and tonic could be unbearably bitter for starters, but it could also induce
vomiting, headaches, nausea, tinnitus and other more dangerous side-
effects. So-called cinchonism, a saturated level of quinine in the blood,
was once the goal of ambitious malaria healers until its harmful side-
effects became more obvious. One medical authority battling 'remittent
fevers' in the West Indies in the 1860s noted with some degree of pride
that of 165 incidences of cinchonism he observed, 'only two proved fatal'.
Not surprisingly, concentrations of quinine drinks would equilibrate
to more modest levels in subsequent decades. It might be noted that
a remarkable number of less savvy drinkers during the twenty-first-
century's COVID-19 pandemic, upon learning of the crude similarity
between chloroquine and quinine, began mixing gin and tonics with
new enthusiasm, hoping that their bitter beverages might afford them
special protection against coronavirus. These thirsty drinkers apparently
cared little about the rather limited molecular overlap of these two
antimalarials, the many contraindications of both substances and the
significant risks associated with cinchonism – along with the fact that it
would take almost 8 litres of tonic at today's maximum legal quinine
concentrations to yield one standard malaria dose.[6]

One therefore realizes that even the amazing Andean 'fever drug'
of quinine can cause an array of negative side-effects, especially when
taken at suboptimal doses. Credited with alleviating much suffering
from malaria around the world, while aiding Europeans in their conquest of tropical lands as they beat back unwelcoming natives who

enjoyed malarial resistance, quinine was also building up a chequered reputation by inducing various unfortunate side-effects in its consumers. Even artemisinin, as one of our day's latest hopes for conquering malaria, can itself produce bouts of diarrhoea, stomach cramps, heart blockages and neurotoxic reactions. More than many other classes of medication, antiparasitics introduced into a human digestive system or bloodstream can pose grave dangers to their taker, even if administered in normal doses, which may only be noticeable after long exposure. The question then becomes how to balance an antiparasitic's costs with its benefits. Is contracting malaria always worse than suffering the collateral effects from the substances we take to prevent it?[7]

Traditional Malaria Remedies

The human experience with folk cures for malaria is rich and deep, with most continents and regions developing their own favourite remedies for protective, curative and palliative effects against this age-old affliction. Since Galen of classical Rome, Western medical experts traditionally relied on purgatives and emetics as well as bloodletting to rid the body of evil humours and noxious fluids. When fevers struck, herbs or other substances that could induce diarrhoea and vomiting were the main drugs of choice across Europe. According to Mark Honigsbaum, before the Spanish travelled to South America to observe the advantages of quinine, their typical malaria remedies might involve 'viper's broth, crab's eyes, and spiders, which – imprisoned in rags and hung around the patient's neck – were thought to act like therapeutic "cocoons," drawing the "poison" of the fever into their own bodies'. Ancient China's own list of anti-fever concoctions numbers in the thousands, and it was only through methodical trial-and-error testing that chemist Tu Youyou was able to single out *qīnghāo* and then in 1972 isolate artemisinin from this plant. One can only assume that China's many alternative remedies to *qīnghāo* usually produced even more disagreeable side-effects than those produced by this herb of choice, with some effects being similar to malaria's own symptoms of nausea, vomiting, appetite loss and dizziness. It seems that in these early treatments against malaria, one typically needed to suffer symptoms like those brought on by the disease before hoping to eliminate those caused by it.[8]

Searching for other holy grail cures, more recent generations of plant hunters have fanned out across continents in search of other lost medicinal remedies, often visiting village folk markets to talk with herbalists and inspect their wares to see if the next cinchona or qīnghāo powder might be found. One source lists that at least 1,200 plant species across the world have been used in treating malaria. A group of ethnopharmacologists combing through villages in five countries of southern Africa, for example, identified eighty different plant species employed in malaria therapies, with most of these originating in Zulu practices. The stunning variety of medicinals they found include barks, leaves, stems, flowers, fruits and roots, taken as infusions of water, salves or powders, or inhaled. Some were used against symptoms, while others were aimed at cure. A few of these medicinals were imported from afar, such as the guava (*Psidium guajava*) used in South Africa's KwaZulu-Natal province, which originated in South America. Biochemists have isolated two of guava's most potent extracts, found in the bark and leaves, and demonstrated their mild abilities to inhibit plasmodia growth in laboratory animals. Yet a follow-up study showed very limited benefits of ingesting them, since modest concentrations of them were found to cause liver inflammation in rats. Complicating this pharmaceutical search is the fact that many traditional antimalarials come as mixtures or are administered at various doses at different frequencies. There is certainly fame – and profit – for the person who can locate the next miraculous antimalarial elixir sitting in the back shelf of a local herbalist, but any future patient who consumes these medications will likely face a variety of discomforts and even risks. *Plasmodium* is a tough protozoan to crack precisely because it inserts itself so seamlessly within a host's body, with a strike against these protozoa cells almost always coming with a strike against human cells.[9]

Italy's own tradition of substances, remedies and rituals employed in the struggle against malaria is likewise very rich. An interview with a Sardinian involved in the post-war DDT campaign revealed that to deal with malaria's fevers, his family relied on such home concoctions as liqueur of genziana and oil of merluzzo (whiting) as well as quinine pills. In the southern region of Calabria, one medicinal survey shows that parts of at least fifty kinds of plant and ten kinds of animal have been commonly used as malaria preventatives (prophylactics) or cura- tives (therapeutics). A malaria carrier's characteristically enlarged spleen was sometimes counteracted with ox gall or goat dung, for example. Bloodletting was also practised here until the mid-twentieth century for ridding a person of recurring fevers. If one slept in the open air, the surrounding bad air might be fended off by moving closer to an open fire. Drinking wine (especially on an empty stomach), chewing tobacco and swallowing the spittle, ingesting ground garlic with vinegar or olive oil mixed with absinthe were also meant to alleviate fevers, or at least alter one's senses for long enough to mask some of malaria's harsher symptoms. Magical and religious rituals likewise seemed to offer promise, and in many instances were certainly less harmful than, say, the alternative of swallowing a pinch of soot or of black pepper (*Piper nigrum*). It seems that praying to Madonna della Febbre, housed in Cosenza's Church of San Domenico, may well have provided more short- and even long-term relief than inhaling any of the aromatic powders sitting inside a medical man's cabinet, while providing at least some condolence for the mind if not for the body.[10]

This broad array of antimalarials is clearly a reflection of the gravity of this malady, and of the intensity of suffering that can be traced to this parasite. But many malarial individuals simply learned to accept that they would need to live with the fevers, which, like other hardships, from winter winds and bedbugs to sore feet and hunger pangs, might be worse in some months than others. Death by malaria was an ever-present threat, too – but mortality rates were always lower than mor- bidity rates, with one estimate suggesting that in Italy in the 1880s, only

2 to 4 per cent of malaria cases proved fatal. A more dramatic malaria remedy that some opted for was to simply flee from a feverous land in search of more salubrious climes. In 1882 Senator Luigi Torelli branded malaria as 'the biggest evil, the principal evil' and the main reason why so many Italians immigrated to America. Denis Mack Smith, one of the doyens of Italian history, judged the peninsula's eventual victory over malaria as probably 'the most important single fact in the whole of modern Italian history'.[11]

Even as quinine was becoming more widely available in Italy across the nineteenth century, and was being mixed and refined in more effective doses, many other remedies against malaria were not so easily discarded. Popular understandings of the disease also convinced villagers to avoid humid and windless areas, obvious sources of the *mal'aria*. The disease's accepted aetiology meant that stagnant water and foetid ponds needed draining or flushing with fresh water to dissipate their noxious effluvia. Eucalyptus seedlings imported from Australia became a health-promoting tree, planted widely across soggy areas to promote drier soils, which would exude *buon'aria* – good air. By 1900 some 35,000 hectares of eucalyptus seedlings were planted across Sicily alone for desiccating land and purifying pestilential airs. Soon, canals and other drainage schemes aided by mechanical pumps were built along coastal wetlands or meandering rivers for 'reclaiming' arable land, in a grand process termed *bonificazioni*, or simply *bonifica* – 'improvement'. Projects for sanitizing, rebuilding and reforesting landscapes were ways to promote hygiene while enhancing agriculture, with the goal of improving human health as well as land productivity. Newly excavated canals and dykes also served to regulate water flow and diminish flooding. The late nineteenth-century discovery by British and Italian scientists that mosquitoes transmitted microscopic parasites between people served to legitimize and reinforce the general strategy of draining pestilential swamps (*bonifica igienica*), expanding farmland (*bonifica agraria*) and controlling floods and improving irrigation (*bonifica idraulica*). By 1911 Italians were battling parasites and mosquito vectors in order to improve human health, boost land productivity and tighten flood control in a systemic effort called *Bonifica Integrale* or Integral Improvement. This would be co-opted and enlarged by Benito Mussolini to become a countrywide social and economic programme that would be admired and emulated by leaders around the world, including Franklin D. Roosevelt, who sent American observers to Italy as part of his government's planning process for the Tennessee Valley Authority and larger New Deal. An explicit goal of these national development programmes was to beat back the malaria parasite by creating healthier and more productive landscapes.[12]

Bonifica Umana – Human Improvement

As part of Italy's mobilization against malaria, government officials would place more confidence in quinine, seeking to make it even cheaper and more widely available. With the disease endemic across whole villages, malariologists recommended population-wide drugging, their goal being to pack quinine into the bloodstream of every individual across a region, first as a cure and then as a preventative. An 1895 law mandated that the price of quinine be regulated across the Kingdom of Italy, leading to its easy distribution in local tobacco shops, with subsequent laws directing that it be manufactured in a government facility at Turin. Quinine was sometimes mixed with sugar or coated with chocolate to make it more palatable, all with the aim of coaxing both young and old to regularly take their doses. So much faith was placed in mass quinine dosing that it was joined with the wider *bonifica* endeavour of managing landscapes. *Bonifica umana* – human improvement – was thus born under the assumption that human bodies could be shaped and chemicalized for the betterment of all. What began as a mass-drugging campaign would be incorporated into the rising movement of eugenics. For advocates of *bonifica umana*, it made sense that human bodies could be improved through selective breeding as well as mass chemicalization.

State-manufactured quinine pills (*Chinino dello Stato*), even when coated in chocolate, continued to produce the side-effect of nausea, so incentives were required to keep people taking their pills. In coming years, quinine would be further subsidized or given out for free, reinforced by advertising campaigns. Private and public workers such as soldiers stationed in malarial zones were soon required to take daily (or biweekly) quinine pills, with inspectors seeing to it that every person actually swallowed their doses – even if some individuals experienced significant adverse reactions to the drug, which included vomiting, diarrhoea and visual or olfactory disturbances that could continue for long after they refused to take their medication.

Italy's quinine mandate certainly brought some relief from malaria, with records showing rates of spleen distension and numbers of febrile patients dropping during the first decades of the century. Yet the medicinal focus of *bonifica umana* was only one of several reasons why malaria numbers declined, with the construction of sturdier houses, screening of windows and draining of wetlands also contributing to *Plasmodium*'s retreat. Once the claims about the mosquito's role in transmitting malaria were more widely accepted, greater resources would be diverted into the struggle against the insect and not just against the parasite.

As the malaria campaign progressed, louder voices protested that the systematic human chemicalization of *bonifica umana* posed distinct drawbacks. Not only did observations reveal that *Plasmodium* was becoming less susceptible to quinine, but Italians were becoming increasingly reluctant to continue taking their doses, especially when they observed the drug's diminishing results. Malariologist Giuseppe Tropeano, working in the southern regions, became rather critical of the mass drugging programme, declaring as early as 1908 that 'malaria parasites are becoming increasingly resistant to quinine,' leading to a general and 'profound disillusion' with widespread drugging: 'it's not right and it's not human to continue deceiving the people with a prophylactic that is not really a prophylactic.' In Italy's marshy Po Valley in the north, many railway workers had stopped taking their quinine doses altogether, some of them through artful deceit, for example by hiding their tablets under their tongue when offering an open-mouthed inspection, and then turning the corner to spit out the payload – or else, with sleight of hand, these unconvinced workers 'substituted the obligatory pill with a tiny bread ball before swallowing it instead'. Historian Frank Snowden reports that public health officials in 1909 estimated that more than half of the distributed quinine was not consumed. Peru's wonder drug was apparently continuing to cause nausea with full effect, but showing diminishing utility against the disease it was meant to combat. Experts also began judging the medication as hazardous for children and pregnant women and 'disastrous even for the strongest stomachs'. No wonder more and more of the people exposed to the disease felt it was better to risk hosting the parasite than to ingest a proven toxin.[13]

Myths die hard, and so did the government recommendation to swallow 40 centigrams of quinine sulphate each day, with quantities doubled if one felt malaria's symptoms setting in. Although quinine would be distributed across India and Nigeria in the following decades to register significant antimalarial successes, *Plasmodium*'s resistance to quinine in Italy as well as in Brazil was becoming well known by the experts. Careful observations showed that cinchona extract simply did not quell malaria outbreaks the way it used to. *Plasmodium*'s intensive exposure to quinine across the Italian peninsula meant that natural selection began producing very resistant strains of the parasite. It may also be true that Italy's methodically organized, state-subsidized quinine distribution programme led to some of the most quinine-resistant *Plasmodium* strains anywhere. The typical response to such drug resistance was to increase the frequency and concentration of quinine doses – which in turn led to still greater parasite resistance.[14]

Sardinia Meets the Pharmaceuticals

With quinine losing its power in Sardinia, there arose new enthusiasm
for earlier folk remedies, as well as heightened searches for new antimalarials. The state-sponsored *Bonifica Umana* was also leading to
health problems that were proving to be worse than the malaria it was
meant to defeat, since quinine's side-effects could become cumulative.
One source mentions that 'traditional cures such as herbaceous concoctions and bloodletting were again being practiced by malaria sufferers, who began hiding from the official doctors when they showed
up or else locked their doors when they saw them coming.' The first
synthetic antimalarials were also finding their way to malaria zones.
After quinine's chemical structure was revealed in the 1920s, the race
was on to produce variations of this complicated molecule, with more
than one chemist hoping that their own novel creation would provide
even better *Plasmodium*-killing properties. Historian L. J. Bruce-Chwatt
reports that some 20,000 compounds would be synthesized in these
decades for use as potential antimalarials, all of which needed screening for positive or negative side-effects. The first such promising synthetic, called 'plasmoquine' (or occasionally 'pamaquine'), appeared
in 1926 and exhibited an excellent ability to kill plasmodia. Werner
Schulemann, the drug's principal discoverer, declared optimistically
that 'the advent of plasmoquine will provide a new and important
weapon' in the struggle against malaria. A few years later, villages in
Sardinia became the sites of some of its first human trials.[15]

For gauging the efficacy and safety of the various compounds flowing from his lab, Schulemann typically relied on injections of these
substances into malaria-infected canaries and then noting any effects.
But other drug laboratories relied mostly on human test subjects, recognizing that what worked on birds did not always work on people. In
a therapeutic coincidence that certainly left more than one individual
dealing with the toxic aftermath of new-fangled chemicals, the new practice of malaria therapy began spreading at the same time that quinine's

powers began to decline. The technique of exposing syphilitic patients to malaria parasites for inducing high fevers and their curative powers depended on the eventual removal of the parasites by an effective anti-malarial, otherwise a patient undergoing malaria therapy would simply be trading syphilis for malaria. So here arose another way of identifying and testing an effective antimalarial, with malaria therapy coming to the rescue of malaria drug development. One might even speculate that the popularity of malaria therapy guaranteed the continued search for effective antimalarials, even if malaria was increasingly limited to tropical regions, far away from the research centres in the wealthier temperate north.[16]

With up to a fifth of all Europeans and North Americans suffering from syphilis in the 1930s, local hospitals, asylums and other psychiatric facilities that housed advanced syphilitic patients were therefore providing convenient proving grounds for testing novel antimalarials. From the perspective of many pharmaceutical researchers, how else were 20,000 potential antimalarials to be vetted? And with few volunteers willing to be subjected to these substances, prisoners and prisoners of war became common subjects – since they required less coaxing, in the twisted ethics of the day. As late as 1948, a special commission was established to determine acceptable ways for convincing prisoners at the Illinois State Penitentiary to take part in antimalarial experimentation, since the typical offer of a pardon or perhaps a reduction in their sentence was not always considered a humane way to gain their cooperation. The commission would recommend in its final report that all potential prison volunteers of such tests should at least 'be adequately informed of the hazards, if any'.[17]

As it happened, the hazards of Schulemann's own plasmoquine 'weapon' were many. One may also wonder to what extent Sardinian villagers were adequately informed of this new drug's hazards, if any. Most initial reports remained enthusiastic about the drug's powers, especially in its ability to combat the parasite. But the problem given short shrift was that plasmoquine produced a range of negative side-effects, especially in higher doses. Following a generally positive review of plas-moquine in 1933, army doctor Henry Dixon felt nonetheless obliged to mention that some of the drug's prescribers took on a 'cyanotic or grey-ish tinge' or complained of abdominal pain, while developing jaundice or perhaps tachycardia. Another expert admitted that the new drug 'may cause mental derangement'. In confronting the shrinking powers of quinine, it seems that these investigators were fast to trumpet the powers of plasmoquine, especially since there were few other effective antimalarial choices. But with accumulating evidence of the dangerous side-effects of this new drug, even if it showed an amazing ability to kill plasmodia, plasmoquine would also fall by the wayside. In all likeli-hood, the parasite did not have time to build up much resistance to the drug. Villagers back in Sardinia, meanwhile, were realizing that they had met their enemy, and it might not be the plasmodia circulating in their veins, but rather the healthcare experts descending on their villages with briefcases filled with perfunctorily tested medicinals. A retrospec-tive WHO assessment of plasmoquine concludes that in the first few years after the drug's debut, 'severe toxic reactions were observed rather frequently.' A modern list of the reported drawbacks of this medication runs to a full page, and includes hallucinations, loss of hearing, ringing in the ears and heart problems, among others.[18]

In continuing this antimalarial saga, the next widespread drug to hit the streets was quinacrine, also known by its tradename Atabrine. This newest pharmaceutical soon replaced Schulemann's discovery during the Second World War. The toxic effects of this bright yellow pill seemed more benign, and a few creative druggists – perhaps unwilling or else unable to get rid of their stocks of quinine and plasmoquine – began producing mixtures of all three drugs to claim the best efficacy. As Carlo Levi described his experiences in southern Italy during the war, 'In the few dust-covered pharmacies of these villages one could never be sure whether a prescription would be accurately compounded or whether it would come out, with good luck, as a mixture of harmless powders.'[19]

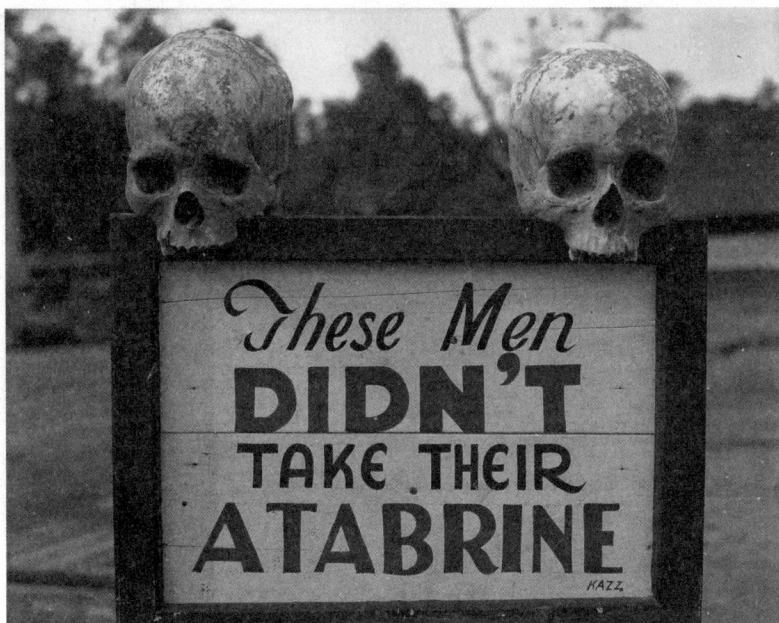

Sign displayed at the u.s. Army base in Papua New Guinea, *c.* 1943.

With ongoing distribution of Atabrine, the yellow pills became integral to an Allied soldier's daily rations. Of course malarial casualties were a serious impediment in the battles of the South Pacific, as well as in those of southern Europe, and the generals in command saw in these 100-milligram pills a way to keep their troops up and charging. But a new battery of side-effects from this latest drug convinced many soldiers to quit swallowing their pills, since many of them began to experience not only nausea, vomiting, diarrhoea and even psychosis, but exfoliative dermatitis and itchy skin bumps called *lichen planus*. Atabrine was also observed to cause foetal deaths in rats, even if one hopeful researcher added that in human users, 'successful pregnancies have been reported.' There was, moreover, the irksome side-effect experienced by at least half of faithful Atabrine consumers of seeing one's skin gain a yellow tinge.[20]

Not surprisingly, Atabrine was soon being peddled in Sardinian pharmacies, or given out for free, first by the government and then by the Allied forces. Even though malariologists began to favour killing mosquitoes over mass drugging, Atabrine would continue to be the

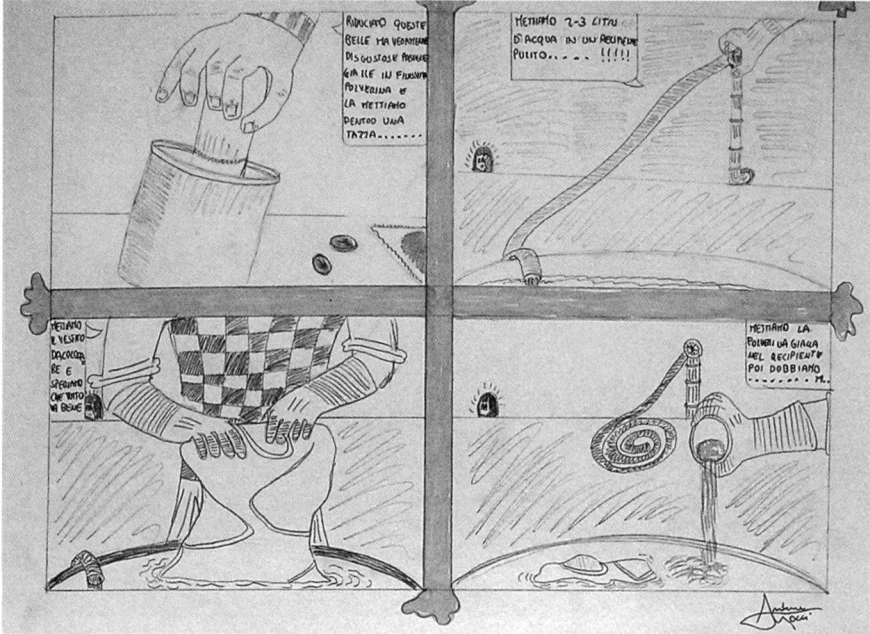

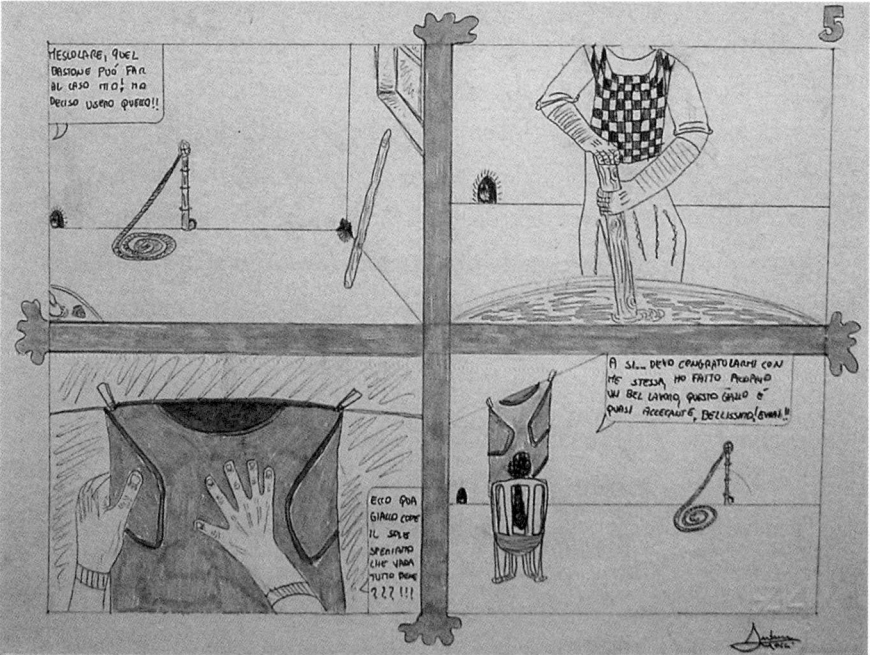

Drawings by a Sardinian primary school student who was told by her grandmother that the Atabrine pills freely distributed in the late 1940s were sometimes ground into yellow clothing dye rather than swallowed for disease control.

official antiparasitic on the island for many years to come. Meanwhile, Sardinians would also start discovering the *non*-therapeutic uses of this latest antimalarial. In an oral history project carried out many years later, which was designed to get Sardinian schoolchildren talking with their grandparents about the post-war DDT campaign in order to gain insights about the insecticide's long-range health effects, Atabrine often found its way into their conversations. The infamous yellow pill proved to be an important subject, not only because it was so widely distributed, but because so many of those grandparents refused to swallow it. In one poignant dialogue, a grandmother recounted to her third-grade enquirer that the yellow pill could be ground into powder and mixed with water to serve as a splendid clothing dye. She further revealed that during village festivals and dances in those post-war years, partygoers who normally came dressed in white shirts and blouses would instead sport striking yellow apparel at these events. Many Sardinians had apparently found better uses for their free Atabrine doses than battling a parasitic disease and suffering the toxic consequences.

Sardinia Meets the Pesticides

Conscientious malariologists began wondering if the chemical warfare being waged in peoples' bloodstreams counterbalanced the pain and suffering of malaria. Such experts also appreciated that maintaining concentrated, population-wide drug concentrations meant that there was no end in sight to such suffering, since malaria would return the moment the pharmaceutical shield was taken down. For such reasons, even more attention was turned to attacking the mosquito rather than the parasite that it carried. Just as important was that the spraying of DDT was showing itself to be spectacularly lethal against insects. The Sardinian Project (1946–50), organized and administered by the Rockefeller Foundation and paid for largely with war reconstruction monies, aimed to eradicate every malaria-carrying mosquito on the island. The Rockefeller Foundation saw in Sardinia the opportunity to test whether an indigenous mosquito could be completely eliminated from an island, with the implication that successful mosquito eradication here might be upscaled to whole continents or even to the globe. DDT would be sprayed by the ton. A 1950 British documentary described the Sardinian Project as a declaration of war, and its 32,000 DDT sprayers as 'indeed an army'.[21]

Yet a good many Sardinians remained unconvinced that killing mosquitoes would replace 'bad air' with 'good air' to finally expunge their island of malaria. While they built their towns far from pestilential swamps or brought their children to the beach to breathe fresh air, they did not easily come to believe that such activities were a way of distancing themselves from the mosquitoes that were transmitting parasites. Or perhaps they did not want to admit that bloodletting and Sunday prayer held little ability to cool their fevers. Many Sardinians did not appreciate the extent to which their lifestyles and landscapes had been fashioned by an invisible parasite carried by a mosquito.

But as soon as the scientific world agreed that mosquitoes were the missing link in the malaria mystery, governments were much more enthusiastic about doing what they could to kill this bloodsucking insect, or at least limit its contact with people. In the years between the world wars, sturdier houses, screened windows, thicker clothing and avoidance of sleeping outside at night all helped. And since mosquitoes in flight were so difficult to kill, the project of killing them as floating larvae became the first line of attack. Draining marshes, straightening and canalizing rivers, filling in ponds and otherwise drying up and planting over marshy areas were the goals of organizations intent on destroying mosquito populations while simultaneously 'reclaiming' farmland and 'training' rivers. Across Sardinia and across the world, thousands of kilometres of waterways and wetlands were channelled or filled in the name of malaria prevention as well as agricultural expansion and flood control – with the little parasite being a prime mover of much of this work. DDT application became another invasive step in the saga to eradicate mosquitoes, relying on powerful toxicity rather than habitat destruction. The face of the planet would never look the same after humans and their activities began reforging the biosphere in an effort to exterminate their protozoan cohabitant. Some argue that the Anthropocene, the geologic era marked by humanity's imprint, began with Watt's steam engine or with the laboratory synthesis of fertilizers, or even with the first atomic bomb. But the Anthropocene may really have started when malaria was traced to the mosquito – and the parasite it carried – putting in motion a vast sculpting of the Earth with innumerable reclamations and reforestations that were then covered with pesticides to collectively forge an indelible mark in the planet's stratigraphy. Call our own human epoch the Age of Mosquito Control.

The rising science of ecology also assisted in humanity's struggle against the mosquito, since a keener understanding of the insect's life cycle, preferred habitats and diet, and reproductive needs and behaviour, all served to identify better ways to kill it. 'Applied entomology' was a term born in the late nineteenth century that reflected the goal of controlling our principal insect enemies, with mosquitoes topping the list. In these early years, such utilitarian entomologists devised any number of ways to disrupt mosquito habitats, going beyond physical draining and dyking to include such desperate techniques as 'oiling' – or the generous spreading of crude oil and other chemical waste across water surfaces. One can only begin to envision the sheen of glimmering petroleum that was flushed across ponds, through reeds and around submerged grasses in hopes of smothering larval mosquitoes. Here the chemical disruption did not take place in human bloodstreams – at least not immediately – but across entire aquatic ecosystems, with enormous varieties of arthropods, from dragonflies and water fleas to damselflies and crayfish, likewise succumbing to the toxic effects of oozing hydrocarbons. Frogs and birds that depended on daphnia or midges, or green and red algae, in turn concentrated the toxins seeping along the riverbanks and estuaries, with the first level of poisoning moving up the food chain. In fact, wetland draining by steam shovels or dykes may have

'Training' (and draining) rivers for malaria control in the
Roman Campagna, Italy, 1930s – before (*below*) and after (*above*).

seemed rather innocuous compared to these purpose-built oil slicks. From our modern perspective, such grizzly scenes of splattered oil demonstrated that the struggle against malaria served to decimate biodiversity, with noxious cocktails of ketones and aldehydes going on to threaten human physiologies, making us wonder again whether malaria treatments were worse than the disease.[22]

Beyond crude oil, other poisonous substances were employed in the battle against disease vectors, whether sprayed across habitats, combusted as repellent or rubbed directly onto human skin. The search for stronger controllers and killers of disease-promoting insects logically borrowed insights from agriculture, since farmers had long been dealing with ways to rid their crops of troublesome critters. A farmer's publication from 1894 catalogued some of the substances thrown at midges, aphids, beetles and their like, including soapy water, camphor, turpentine, sulphur and pyrethrum – if one could afford them. That some of these substances could be dangerous to human beings was also well known. Four decades earlier, an Italian almanac had cautioned about the indiscriminate use of such insect deterrents. 'One can never be too careful, since the smallest miscalculations can end up being fatal,' it read. But it seemed that the accumulated wisdom of farmers was not always passed on to healthcare workers. Chemical substances sprayed indiscriminately across marshes or applied copiously to the body ushered many a malaria sufferer to the cemetery. As in the quest to find better parasiticides, humanity's search for stronger insecticides could mean two steps backwards for every step forwards. Early safety tests with DDT proved to be 'somewhat alarming', even though they were never publicized. Edmund Russell uncovered some preliminary studies about DDT that revealed important doubts about the newly discovered chemical: 'When eaten in relatively large amounts by guinea pigs, rabbits and other laboratory animals, [DDT] caused nervousness, convulsions or death depending on the size of the dose.' Not for another two decades and the publication of Rachel Carson's *Silent Spring* in 1962 would this and other pesticides be widely exposed as substances that could seriously poison humans and other living organisms indirectly and invisibly over short and long spans of time.[23]

Pyrethrum was perhaps the safest deterrent against Sardinia's mosquitoes, since this naturally occurring substance seemed harmless to people. However, the high cost of the powder and its unpredictable supply meant that this insecticide would not be applied across very many of the island's marshes and river banks, instead serving mostly as an insect shield for the few. It was produced by drying and powdering chrysanthemum flowers, a practice developed in ancient China and the Near East, which attracted the attention of modern chemists, who were eventually able to reveal the flower's active ingredients but not synthesize them in the lab. Instead, the chemical industry began producing 'pyrethroids' that were derivatives of the flower's original ingredients but were much more dangerous. Even the organic extracts of chrysanthemum that are still used today by mosquito-avoiders are neurotoxins at any dose and are linked to various cardiovascular disorders. It seems that a chemical for killing mosquitoes that is easily synthesized or easily collected without harming people may not exist.[24]

Despite such challenges, a few relatively cheap insecticides became widely used across the twentieth century, sometimes as offshoots of the dye industry. These contained arsenic and were labelled with such cosmopolitan names as London Purple and Paris Green. Such compounds

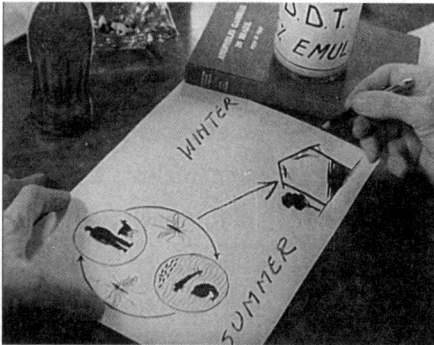

Scenes from the film *The Sardinian Project* (1949, dir. J. D. Chambers), including unloading barrels of DDT, which was carried out under the auspices of the United Nations Relief and Rehabilitation Administration and Rockefeller Foundation, 1946–50.

were dusted over crops as early as the 1870s to deter beetles and sprayed across marshes and riverways in the 1920s to control mosquito larvae, soon finding their way to Sardinia to represent the first European trials of these compounds against the larvae. Although there are no records of human poisonings in Sardinia's villages of Posada and Portotorres following those initial Paris Green dustings, one can be sure that more than one villager felt sick to her stomach a few days after the arsenite powders settled to the ground. Paris Green is an unambiguous human toxin. Even if these arsenic pesticides lost favour in coming decades, in large measure due to the discovery of the more efficient killing properties of DDT, they continued to be utilized by insect exterminators who saw special advantages in their ease of use and voluminous availability. Also useful as a pigment in green paint, Paris Green would be banned outright in the 1960s after its extreme toxicity became more evident. There was, for example, the mysterious illness of Clare Boothe Luce, U.S. ambassador to Italy in the 1950s, which was attributed to the pigments of Paris Green impregnated in the wallpaper of her Roman villa. For such reasons, DDT was welcomed as the insecticide of choice in Sardinia's post-war anti-mosquito campaign. There, across marshes and along stream beds, over hillsides and inside pastures, goat sheds and bedrooms, it was layered over the residue of Paris Green, itself layered on top of petroleum, which may have already been dusted with pyrethrum or sprinkled with turpentine. Toxins kept accumulating as mosquitoes kept buzzing.[25]

DDT's half-life, or the time required for half of the chemical to disintegrate, varies from two to fifteen years in soil and up to 150 years in water. To glance at the period-piece photographs of Cagliari's main harbour displaying rows and rows of 55-gallon barrels, all marked 'DDT 25%', which would be distributed over the island to be further diluted and sprayed across fields, riverways, shores and dwellings, is to realize that there is still plenty of DDT lurking in Sardinia today.

The island's massive DDT programme, which is generally celebrated as the critical event that finally cleared Sardinia's pathway to modernity, can also be viewed as just another stage in the ongoing toxification of the island and its inhabitants. In the grand project of chasing away malaria parasites, Sardinians' internal organs have been flushed with quinine, plasmoquine and Atabrine, and joined through their skin and lungs with pyrethrum, petroleum, Paris Green and DDT. With DDT's mosquito-killing power waning by the 1950s, due to growing insecticide resistance, still other chemical arsenals would be deployed, and other medicinals would be swallowed. Noted ecologist Marston Bates, an observer and sceptic of the Sardinia Project, cautioned that DDT was a 'sledge hammer approach to mosquito control' since this potent spray could inflict so much secondary damage to surrounding organisms and natural systems. There is also little doubt that many antimalarials ingested by Sardinians over those years likewise acted as sledge hammers in human bloodstreams.[26]

One wonders what an analytical chemist would find if sifting through Sardinia's soils and effluents – where a medley of synthetic compounds has been deposited and come to rest. One public health project aimed at tabulating the causes of death among Sardinia's former DDT sprayers revealed that their rates of cancer were actually not much higher than those of their neighbours who did not work on the front lines of the DDT campaign. Pierluigi Cocco, the epidemiologist in charge of the study, concluded that 'we found little evidence for a link between occupational exposure to DDT and mortality from any of the cancers previously suggested.' But with almost everyone on the island in those days being exposed to blow-by pesticides, whether in the stables and fields, or in their homes and offices, while swallowing a cornucopia of pharmaceuticals, it seems that toxin exposure may be a general hazard of living in Sardinia, with toxin-induced cancers being a health problem faced by all islanders regardless of whether they held a DDT spray gun. Such a view is supported by recent government tabulations which show that Sardinia as a whole suffers significantly higher tumour rates than other Italian regions.[27]

One might also conjecture that the chemical warriors in charge of the Sardinian Project did not pay sufficient attention to ecology.

It seems that classic works such as Lewis Hackett's *Malaria in Europe: An Ecological Study* (1937), published almost a decade before the DDT sprayers began crisscrossing the island, were mostly shelved once the power of the new insecticide made itself known. Medical doctors in the Rockefeller Foundation overseeing the project generally substituted stethoscopes for spray guns, seeing more value in administering pesticides than parasiticides. The engineers in charge likewise focused their efforts on developing better spray technologies or insecticide mixtures, for instance, and not ecological understanding of the parasite and its carrier. They relied instead on so-called DDT bombs, for instance, which were fashioned out of clumps of seagrass saturated with insecticide and lowered into water wells suspected of harbouring mosquito larvae. One anecdote relates how occasional dedicated sprayers would go beyond the call of duty to win over DDT sceptics still unconvinced of the benefits of the newfangled pesticide by actually swallowing samples of the pesticide in front of their perplexed onlookers, who were protesting that their wells should be left uncontaminated. That such exhibitionists did not keel over on the spot after performing their parlour tricks was probably due more to DDT's limited solubility in water (and so, in water-saturated human tissues) than to its supposed safety. In fact a host of Sardinian apiarists and fish farmers registered fiery complaints after watching the numbers of bees and fish plummet following dousings by the spray crews. Certainly mosquito numbers also plummeted across the island, at least during the first years before DDT resistance set in. Local mosquito abatement offices that took over after the all-out spray campaign ended in 1950 would complain about the inferior potency of 'Italian' DDT when compared to earlier results using 'American' DDT. In subsequent decades, mosquito numbers would march back up to previous numbers despite the rising concentrations of DDT used in insecticide sprays. In the end, one might speculate that the Sardinian people themselves, like the mosquitoes now swarming the island, may be some of the most DDT resistant on the planet.[28]

Brave Medicinal World

So accustomed were Sardinians to the steady stream of chemicals being pushed at them in the name of public health that many islanders simply began referring to any of these substances as *la medicina* – the medicine. For these consumers, *la medicina* was ingested in the form of quinine pills, plasmoquine pills and yellow Atabrine pills – especially when they were coerced to do so. But *la medicina* was also the name given to the substances being sprayed in and around riverways, sheep stalls, courtyards, pathways, bedrooms and clothing as a killing agent for pesky insects. *La medicina* used in the immediate post-war era, which experts labelled DDT, was also occasionally ingested for persuasive effect, as seen previously. But with so many different kinds of insecticide and pill, this entire battery of chemicals simply merged in the popular view, with all supposedly useful for improving human health. For the casual onlooker, the exact target or methodology of these substances did not matter much since most of them would find their way inside the body anyway, and so might as well all be medicines. Many threatening parasites exposed to such medicines might perish, as might the insect transmitters of these parasites, but a remainder of such creatures always managed to resist this chemical barrage, just as traces of the 'medicine' typically remained in the human body – sometimes for decades.[29]

In the years after the Sardinian Project, with DDT still being sprayed across the island and around the world, there arose a steady stream of even newer and hopefully improved mosquito killers. These included substances with names like malathion, naled, permethrin, prallethrin and etofenprox, together with mosquito repellents such as DEET, Picaridin, IR3535 and many, many others. Live organisms were also pitted against mosquitoes, with one of the first being *gambusia*, the invasive mosquitofish that was propagated in waterways across Sardinia and the world during the interwar years. Copepods and fungi would become the mosquito biocontrols of choice in the 1970s,

IL D.D.T. NON E' VELENOSO

L A recente scoperta del D.D.T. ha rappre-
sentato il colpo di grazia per la specie
anofelica: quando, soprattutto, si tenga
conto dell'innocuità del prodotto nei ri-
guardi dell'uomo, dell'acqua potabile e del
bestiame, purchè usato a bassi titoli per-
centuali. Esperimenti praticati sù pecore
sarde, costrette a bere acqua ricoperta da
un velo di D.D.T.-nafta, hanno dimostrato che alle pecore
stesse non arreca danno la soluzione in nafta del D.D.T.
prevista come massima per il trattamento antilarvale, e nem-
meno a dosi triple di questa.

'DDT IS NOT POISONOUS' announced a 1948 newsletter of ERLAAS, the agency
responsible for carrying out Sardinia's DDT spray campaign.

eventually joined by bacteria such as *Wolbachia* – with all of them
disrupting different stages of a mosquito's life cycle. Thereafter, toxin-
producing bacteria called Bti (*Bacillus thuringiensis israelensis*),
described as a 'living' insecticide that could be sprayed over marshes
to control mosquitoes, came into widespread use, even if there were
claims that it upset ecosystems. Eventually, the release of genetically
modified mosquitoes became the high-technology tool to join the list
of ways for controlling and eliminating mosquitoes.[30]

Joining this march of mosquito killers and repellents came the lit-
any of substances designed to kill or repel the creatures that mosquitoes
transfer to us. With the rise and fall of quinine, Atabrine and chloroquine
(and hydroxychloroquine), pharmaceutical companies developed or
refined such antiparasitics as tafenoquine, doxycycline, mefloquine,
atovaquone, artemisinin and primaquine, which, under various trade
and scientific names, were supposed to offer still greater convenience,
heightened potency or fewer contraindications. Yet all such medicines
presented discrete disadvantages, as in the case of primaquine, for
instance, which could damage red blood cells, especially in people with
favism. And with so many people in malarial and ex-malarial areas prone

Typical well-stocked 19th-century German pharmacy.

to favism, as is the case in Sardinia, primaquine would not seem a very useful antimalarial on this island.[31]

Rome's Institute of Public Health explains on its webpage that most medicines now used in *preventing* malaria are different from those used in *treating* it. The U.S. Centers for Disease Control adds that 'No anti-malarial drug is 100% protective.' Meanwhile, the quest for developing a malaria vaccine has proven mostly elusive due to *Plasmodium*'s enormous complexity as a double-hosted parasite, its many developmental stages, its proclivity to lie dormant for long periods and its ability to rapidly evolve. Even the latest malaria vaccine, Mosquirix, which was introduced in 2021 with much fanfare, is marginally effective since it offers only a 30 per cent success rate and must be administered multiple times during infancy, with little efficacy for older children and adults.

If one reflects for a moment on humanity's age-old medicinal war with mosquitoes and the microbes they carry, while considering the plethora of chemicals ingested and absorbed in our fight against them, one may well envision a sort of *Lullaby Spring*, Damien Hirst's pill-case installation that sits in a Munich museum. Hirst's auditorium-sized medicine cabinet is crisscrossed with glass shelves from floor to ceiling for the display of row after row of tablets, pills, lozenges and capsules – some 7,000 in total – arranged according to hue and size to form a rainbow of medicines. On display is a sampling of the combined output of four generations of pharmaceutical laboratories, with many of these pills being shinier counterparts of yesteryear's medicinal concoctions. One cannot help but wonder what the cumulative effects this pharmacopeia of chemicals has had on human cells, on our microfauna and microflora, on our landscapes and mindscapes: sprayed into the atmosphere and scattered over fields and ponds, digested by gastrointestinal tracts and rubbed onto skin – to finally be excreted, secreted and reabsorbed. Some future alien visiting our planet will surely be able to detect this medicinal age in the Earth's strata, making this being-from-afar wonder why Earth's former inhabitants spent so much ingenuity manufacturing and depositing this toxic layer.[32]

Plasmodium and Blood Pressure: A Physiological View

There is no question that many of the pharmaceuticals that we have introduced into our bodies during our infancy and our adulthood are crucial for us, vital for reinforcing antibodies, alleviating pain, speeding recovery from injuries and controlling virulent pathogens. But it is our parasites and not our pathogens that we must be most careful about upsetting with the many chemicals that we expose ourselves to. And some of the organisms that we believe to be pathogenic are actually mutualistic, providing us with benefits that we are only beginning to discover. Parasites are those creatures that we assume are harming us, but upon new information, we find that many provide us with advantages or benefits that our powerful medications are compromising. A large portion of the creatures living in or on us have evolved with us and are now as dependent on us as we are on them.

So it is still important try to identify possible benefits – whether physiological or cultural – that we humans may be reaping from our co-creatures, especially ones such as the malaria parasite that has been with us so long. Surely, 10,000 human generations of living with *Plasmodium* would seem to indicate that this parasite causes us nothing but pain and anguish. But there are many puzzles that are hard to explain. There is the role of *Plasmodium* in malaria therapy, for example, which relies on this parasite to induce strong fevers to alleviate the ills of syphilis, with one disease fighting the symptoms of another. There is the fact that some *Plasmodium* strains may substitute or displace other, more virulent *Plasmodium* strains, as is the case when *P. vivax* serves to dilute or diminish *P. falciparum* in our bloodstream. One can point to the advantages that *Plasmodium*-resistant peoples enjoy when inhabiting or colonizing places occupied by *Plasmodium*-sensitive peoples, with the parasite serving to heighten the fitness of resistant hosts. Likewise, there are the competitive advantages that malaria parasites confer upon individuals who carry sickle-cell trait or various thalassemias and who live in malarial environments, since such individuals are better adapted to co-existing with the parasite than those who do not exhibit these genetic traits. A seemingly simple and dangerous pathogen, *Plasmodium* is, on greater reflection, a cohabitant that can provide its human host with a range of hidden and indirect benefits.

Yet another surprising benefit that *Plasmodium* may be conferring upon us stems from this parasite's effects on blood pressure. Experience shows that beyond malaria's terrible symptoms of producing throbbing headaches or bone-chattering fevers, the disease often produces a general state of lethargy in its human host, typically manifested by lowered blood pressure. It turns out that *hypo*tension is a telltale sign of serious malaria, measured by a systolic blood pressure of less than 80 mmHg in adults, and less than 50 mmHg in children. Some researchers suggest that a malaria person's lethargic state is the *Plasmodium*'s way to calm down and immobilize its host so as to facilitate a mosquito's goal of taking a blood meal and passing on itself to the next human host. Yet it should also be pointed out that experiencing low blood pressure without other symptoms is rarely a serious health concern, except perhaps in the elderly. Instead, it is high blood pressure that is an important health threat, with *hyper*tension now worsening in many countries today, and labelled the 'silent killer' that requires lifestyle changes and prescription drugs. Large swathes of humanity suffer from high blood pressure. Many Africans and those of African heritage are one such group threatened disproportionately by hypertension, as are peoples from other malarial or ex-malarial areas, such as the Mediterranean – for there is a strong correlation between malarial and ex-malarial areas and high blood pressure.[33]

It turns out that the worst, most deadly episodes of malaria typically kill a person via cardiovascular complications in the brain, especially through blood clots forming in cerebral capillaries and composed of *Plasmodium*–blood cell aggregates. Physicians sometimes prescribe medications to further lower a severely malarial patient's blood pressure in order to relax vessel walls and avert dangerous capillary clotting. Biochemical studies reveal that typical plasmodia are themselves able to set in motion a chain of chemical reactions for lowering a person's blood pressure. Indeed, one can speculate here that adaptive strains of plasmodia working across evolutionary time have been able to lower human blood pressure so as to make themselves less deadly to their host, and so less deadly to themselves. In this way, *Plasmodium* can act as a kind of high-blood pressure remedy for keeping most dangerous blood clots at bay, and (most of the time) averting fatal consequences for the host. One can thus realize that malaria parasites produce fever, headaches and lethargy in their host, making the person an easy target for hungry mosquitoes – but as symptoms become more serious, the same parasites work hard to lower a host's blood pressure to protect it against dangerous capillary clotting, else they endanger the host and so themselves. This image of *Plasmodium* manipulating a person's physiology with marionette-like strings to produce a lethargic, feverish stupor is not a flattering picture. Yet such details can help us realize that humanity's relationship with its ubiquitous parasite is clearly much more complicated than that of a simple pathogen that cares little about a person's health. As devastating as the scourge of malaria is, the parasite responsible for this disease also aims to limit the harm it inflicts. As Hans Zinsser offered, 'infectious diseases are not static conditions, but depend upon a constantly changing relationship between parasite and invaded species.'[34]

And now that malaria is finally receding across our modern world, with its annual victims dropping from millions a few decades ago to fewer than a half million today, one must ask what is becoming of *Plasmodium*'s blood-pressure manipulation. Since increasing numbers of people are no longer exposed to malaria – in temperate and tropical latitudes – is their blood pressure now marching dangerously higher due, in part, to their novel *inexposure* to *Plasmodium*? Just as sickle-cell trait and thalassemias no longer confer advantages to their carriers in malaria-free environments, perhaps blood pressures are becoming problematically higher in places where *Plasmodium* no longer reigns. There is also the laboratory observation that mice infected with trypanosome, a protozoa similar to the ones that cause sleeping sickness in humans, actually grow heavier than mice not infested with this parasite. Other observations show that tsetse flies heavily infected with trypanosomes actually live longer, as do estuarine snails infested with trematodes, along with white-footed mice burdened by ticks and botflies. The amazing conclusion here is that if these mice and flies and snails lose their parasites, they will start living skinnier, shorter lives. By implication, we might ask if there are also physiological drawbacks for humans when they lose their *Plasmodium* parasites.[35]

Before drawing wider conclusions, one must acknowledge that human blood pressure is governed by an enormously complex range of factors that depend on lifestyle, age, gender, diet, emotional state, environment and genes – at least seventy such factors, according to one study. Since blood pressure is such a crucial proxy for human health, and is often used to predict cardiovascular concerns, it has been studied extensively. At first blush, then, linking causes of blood pressure to any single factor, including malaria exposure, as well as ancestral malaria exposure, is too facile.[36]

Yet against the backdrop of blood pressure's complexity, one must also keep in mind humanity's age-old experience with its malaria parasite. Viewed from the perspective of millennia, *Homo sapiens* and its *Plasmodium* are an almost indistinguishable pair of interacting organisms, with one partner rarely being separated from the other across vast expanses of human history. Following some 200,000 years of co-existence, humanity has been as unable to shake off its malaria parasite as this parasite has been unable – or unwilling – to kill off its host. From the eye of an alien coming to study life on Earth, a human being is primarily a complicated bundle of cells of which 10 to 40 per cent (depending on the study) carry human DNA, with the rest carrying DNA from non-human creatures such as viruses, bacteria, fungi, archaea and protozoa – including *Plasmodium*. Owing to *Plasmodium*'s tenacious and persistent cohabitation with people, it would be surprising to find that this parasite did *not* play some role in helping steer the mother ship. From the perspective of deep time, it can only make sense that the ubiquitous, micron-long malaria parasite does indeed influence some of its host's physiological processes, even crucial ones such as blood pressure.

When comparing global maps of peoples of malarial heritage with those who suffer from high blood pressure, one can observe a significant overlap: that is to say, many peoples who currently live malaria-free lives, but whose ancestors lived with endemic malaria, now exhibit especially high rates of hypertension. *Plasmodium*'s disappearance could therefore mean that people of malaria heritage are no longer reaping the benefits of having their blood pressure lowered by the parasite. But is it really possible that today's growing epidemic of hypertension can be linked to historic victories against malaria? To remember the case of Sardinia, one needs to wonder whether it is the islanders' past experience with malaria that contributes in some way to current high numbers of long-lived Sardinians. As two scientists have already pointed out, malaria and hypertension may be a 'co-evolutionary adaptation'. Parasitologist Robert Desowitz makes the same observation, adding that 'no one knows how malaria and other chronic infectious disease lower blood pressure . . . . the malaria parasite may be prophylactic, for it not only lowers blood pressure but also appears to decrease the cholesterol level.' Now that malaria has disappeared from Sardinia, it may be more than speculation to wonder whether the island's remarkable longevity will become a thing of the past.[37]

Another clue comes from Africa during the COVID-19 pandemic. It turns out that much of Africa remained remarkably free from coronavirus cases throughout the pandemic; or, at least, most African nations registered extremely low levels of coronavirus morbidity and mortality. In fact, much of the heart of tropical Africa, including East Africa and countries in and around the Congo Basin, reported no COVID-19 cases at all for months on end during the height of the cruellest months of the pandemic elsewhere. That Africans routinely confront a variety of dangerous infectious agents is well known: however, with the epicentre of COVID resistance being located at the core of the world's most malarial region, there are now speculations that malaria itself can help guard against coronavirus. Barring measurement errors or misreporting problems, the search is on to determine whether or how the *Plasmodium* parasite might provide humans with a shield against the coronavirus. One team of immunologists has concluded that 'we hypothesize that the immunological memory against *P. falciparum* merozoites primes SARS-CoV-2 infected cells for early phagocytosis, hence protecting individuals with a recent *P. falciparum* infection against COVID-19 infection or severity.' Other researchers assert that malaria may promote B cell and T cell immunity against COVID-19. But whatever the biochemical pathways, the refrain here is that certain infectious agents may counter other infectious agents, a message we have already encountered. In the case of *Plasmodium*, this parasite may be helping defend people from pathogens as serious as those as SARS-CoV-2. Through the wonders of cross-immunity, a parasite that can kill can also prevent killings. In

Africa as in Sardinia, will the eradication of one microbe unleash the cruelties of another?[38]

Celebrate Bugs that Are Close

As the COVID-19 pandemic ground on through the summer of 2022, with graphs of the disease's mortality and morbidity still showing nearly as many spikes as troughs, and with the world registering more than 6 million deaths since the coronavirus had first been identified two-and-a-half years earlier, it was becoming frustratingly obvious that the virus was here to stay. Vaccines and treatments for the disease were making progress, as were the immune defences of people who were contracting the virus, even as the virus kept mutating, still sickening people with wave after wave of novel variants. It seemed that humanity's immunological and pharmaceutical defences were in an arms race with the SARS-CoV-2 virus, with each new human tactic being met with a viral counter tactic. Yet there remained hope that humans and virus could someday peacefully co-exist, or possibly reach a stage of détente, whereby the costs of maintaining vigilance would convince both sides to de-escalate their tensions, drop their defences and march towards a stable and mutual peace.

Or at least, this is the story that some analysts have been telling in the aftermath of COVID-19 panic, conjecturing that on some future, happier day, a period of greater COVID harmony will reign. Coronavirus and *Homo sapiens*, they assume, will begin experiencing less virulence, with the virus evolving strains that are no longer so devastating. If the virus killed too many humans, after all, it would have nowhere to go. Less than a year after the coronavirus was discovered, it was pointed out that the original strain had disappeared: 'You can hardly even sample the [original] Wuhan virus anymore,' declared one analyst; 'The D614G strain is now the pandemic.' But other scientists cautioned against this facile view, explaining that the coronavirus might be innocuous to some people but deadly to others: 'the virus can be nearly avirulent if it fails to exploit human cells, but it can also be extremely virulent if it causes massive immunopathology, *for example* via the activation of cytokine storms.' Some also emphasized that Theobald Smith's Law of Declining Virulence had been discredited decades earlier, proven to be too simplistic, too deterministic and too naively optimistic when field-checking infections with a range of other virulent microbes.[39]

A few of these sceptics went on to point out that the coronavirus strains had, by some measures, actually become more virulent over time, with any gains in dealing with the virus reflecting human ingenuity in immunology labs rather than viral beneficence. Well into the pandemic, fewer people were catching the virus, but those who did experienced higher mortality rates than two years earlier. Other tabulations indicated that higher mortality rates might be due to the hosts' heavier viral loads and longer viral infections, which both help the virus achieve its main goal of getting itself multiplied and re-transmitted. Making people really ill over a longer period is generally a better viral strategy for trans-mission, at least during the first few years of a pandemic. But in all of these accounts, the virus–host relationship was being defined largely from the perspective of the host and its susceptibility, rather than from the view of the microbes that the host carried. Chapter Five turns harder into this issue of how a parasite may be viewing the world.[40]

But predictions of humanity's long-term relationships with the coronavirus are not all bleak. Some scientists believe that this microbe might learn how to diminish its virulence and co-exist with people, ultimately vindicating Theobald Smith by evolving into forms 'that tend to be less virulent and cause less symptomatic infections'. Other experts believe that this pathogen is transitioning into a simple nui-sance, like a seasonal flu or chest cold, which will rise and fall according to viral idiosyncrasies and a person's immune responses. There may be a day when a coronavirus will be able to live peacefully within most human bodies without causing harm, finally co-existing by a combina-tion of lesser viral virulence and greater human tolerance.[41] Or, as René Dubos wrote more than a half-century ago:

The concrete facts of microbiological sciences have been on the whole easy to discover and their understanding presents no abstruse problem. The real difficulty has been rather to explain why so many varieties of microorganisms, endowed with the ability to kill, usually produce only self-limiting disease processes and often cause no discernible harm even though they persist in the body.[42]

We can step back at this point and wonder what would be *lost* if we finally learned to live peacefully with this virus, or if we actually exterminated this notorious pathogen – or invented a permanent vaccine against it. We might also ponder what has been lost by coming to peaceful terms with history's other virulent microbes, or else by harnessing the modern medical practices and sanitary methods that have dramatically controlled mumps, plague, rubella and typhus. Although eradicating smallpox and rinderpest from the world were amazing achievements, is there anything to be regretted from the loss of these diseases and their associated pathogens? Since the tenets of nature conservation teach us to avoid ushering any species to its extinction, perhaps we are in error when we are finally successful at eradicating a harmful pathogen.

Such a thought experiment can begin to reveal the larger roles of infectious disease, both in the biological realm and in the human one. As agents of illness and causes of death, infectious microbes are constantly lodging themselves in plants, animals and people to produce ecological, evolutionary and cultural responses. For millennia, the terrible diseases of yellow fever, smallpox and malaria have killed and maimed people – but some of those who caught the contagions showed no ill effects, or else went on to recover. Survivors of these and other diseases demonstrate that there is always some accommodation between microbe and host, for not every encounter proved fatal to both organisms. Hosts and parasites have adapted to each other, partially or fully, directly or indirectly, proximally or distally, else they would have gone their separate ways. Disease agents are purveyors of new relation-ships, some of which will become mutualistic rather than antagonistic. Pathogens and parasites present opportunities for new beginnings, and if symbiogenesis is happening all around us, these transformed microbes present new fountainheads of evolution. The same creatures that can cause disease and death can also provide the building blocks of life.[43]

Certainly for anyone who hosts or will host virulent coronavirus or malaria parasites, there is little doubt that such creatures can inflict fever, pain, trauma – and worse. But every human host might also real-ize that these bodily inhabitants can also supplant more serious patho-gens, vanquish enemies and possibly promote a human body's other vital functions or benefits still uncharted and unknown to us. In a key-note speech, noted environmental historian William Cronon once counselled that in tracking the role of nature in human affairs, a crucial insight has been to 'Beware of bugs that come from afar.' Transported epidemics, such as those brewed in Europe and Africa and shipped to the far corners of the colonial world, he explained, served to subjugate peoples whose immunities were no match for the microbes imported by their immuno-resistant colonizers. To this understanding, we can now add another key insight that might be formulated as 'Celebrate bugs that are close.' Within and on us, we are carrying creatures that have provided our bodies and our ancestors with a compendium of life-improving and lifesaving services. Many of our co-travellers are mutualists and not just pathogens, and many of the creatures that we label as parasites and even as pathogens act as our allies by providing us with a range of benefits, some of which we could not live without – or could not live for as long without. Eradicate our parasites, and we may be eradicating the hidden benefits they can bring us.[44]

5

Thinking Like a Parasite

Man is too prone to look upon all nature through egocentric eyes.
To the louse, we are the dreaded emissaries of death.

HANS ZINSSER, *Rats, Lice and History* (1934)

William McNeill's 1976 classic, *Plagues and Peoples*, retells world history from the perspective of infectious disease and all its ramifications, with human societies from prehistory to the present building their lives and lifestyles around the demands of debilitating and deadly microbes. No other historical work had been able to grapple quite as well with how the complexities of epidemiology and immunology shaped the rise and fall of civilizations, especially through the agency of contagious humans migrating across continents and oceans. McNeill's explanations struck a chord with a variety of audiences, from biologists and medical scientists to anthropologists and historians. Decades after its appearance, admirers were still praising the book as a 'landmark' and 'groundbreaking' work.[1]

In McNeill's synthetic narrative, parasites play an important role. Spreading around the globe and across the centuries, creatures that live in and on people and their domesticated organisms have been large and small, persistent and temporary, but always taking without giving – consuming without compensating. Intriguingly, McNeill distinguishes microparasites, which he sees as composed mostly of invisible or tiny microbes, from macroparasites, which consist not only of larger beings such as lions and tigers, but especially of people, who can all threaten their hosts' prosperity and happiness, especially through predation, even thievery, deceit and coercion. These larger parasites bring social and political consequences. In McNeill's telling, even medieval serfdom and colonial conquests, for example, are manifestations of macroparasitism in which poor or defenceless peoples have succumbed to the exploits of human oppressors who came to depend on the surpluses of the hardworking masses. So, too, have warlords, Roman soldiers and conquistadors marched into foreign lands to 'parasitize' hapless villagers or indigenous peoples who struggled to survive their cruel subjugators. Yet macroparasites could also act prudently, as by 'taking only a part of the harvest from subjected communities, leaving enough behind to allow the plundered community to survive indefinitely, year after year'. McNeill painted a world filled with biological as well as social and even metaphorical parasites, in which individuals and communities dealt with not only internal threats such as smallpox, tapeworms and malaria, but external threats including human oppressors and conquerors.[2]

Despite the enormous success of the book, and the many accolades bestowed upon its innovative interpretations of disease in human history, McNeill's distinction between micro- and macroparasitism did not find much traction in subsequent scholarship. There was apparently an unwillingness to take his broad leap of viewing people in fully parasitic roles, which for many readers exaggerated the extent to which ecological and evolutionary ideas could be applied to human societies. At this point in journeying through the parasitic world, we need to revisit this idea of macroparasitism, and the ways in which host and parasite can be envisioned at various scales for illuminating how people interact with other organisms, including other people. In the next pages, I focus on a parasite's eye view of the world, endeavouring to envision things not as we hosts see our bodily inhabitants, but as they see us. This reversal of perspectives helps us appreciate just how dependent we are on our squirming creatures and, like McNeill's macroparasites, suggests how we humans may be taking more than we are receiving – even as we realize the need to compensate the other hosts around us. Can thinking like a parasite clarify the benefits of being parasitized, and the dangers of eradicating creatures that live on other creatures? What can a parasite's eye view add to the explanations of William McNeill or Alfred Crosby, or to their successful popularizer, Jared Diamond, who all envisioned a Columbian Exchange of creatures that went on to ravage distant peoples and revolutionize world history?

The Parasite's Dilemma

In the act of shrinking oneself down to the level of parasites to try to stare out of the eyes of these intimate organisms, it becomes clear that one of their crucial interests is to ensure that their host survives, at least in some ultimate sense. In fact, every single being that lives in or on another being hopes that their host will survive and thrive into the foreseeable future – with the parasite always agonizing over just how far it can exploit the host. Even simple pathogens and not just parasites want to make sure that their host ultimately survives, in place or in time, else their household will disappear, and they perish. As McNeill explains, 'A disease organism that kills its host quickly creates a crisis for itself, since a new host must somehow be found often enough, and soon enough, to keep its own chain of generations going.' Alfred Crosby similarly offered that 'ecological stability did tend to create a crude kind of mutual toleration between human host and parasite. Most Europeans, for instance, survived measles and tuberculosis, and most West Africans survived yellow fever and malaria.' One can also realize that under natural conditions, malaria's *Plasmodium vivax* cannot survive without us, and even the bubonic plague's *Yersinia pestis* cannot keep reproducing, making us sick and even killing us, if there are no human and other animal hosts at some point to infect and keep these bacteria's life cycles going. These tiny beings that can bring people so much pain and suffering actually hold us in highest regard. Like the Cheyenne hunter who ceremonially reveres the bison that he will then hunt and kill, our virulent microbes revere us – or revere us by any other name – since we are crucial to perpetuating them. Our bodily microbes may be experiencing little dilemma in deciding whether to spare or exploit us, being unabashedly enamoured of us, willing to kill for us, or even kill us, in their dedication to keeping our species and so their species alive.[3]

Group dynamics are fundamental in any parasite–host relationship, since sacrificing an individual host's life may offer advantages to other hosts, now or in the future. The logic of kin selection means that an individual's actions may only make sense when considered from the perspective of what can be gained by the individual's relatives, and even by the whole species. Yet a parasite's viewpoints still need scrutiny, for a fraction of these creatures may be acting too virulently for their and their own species' good. Natural selection means that highly virulent individual parasites will generally not be as successful at producing as many offspring as their less virulent cousins. Even anthrax bacteria, the respired spores of which can kill a person in a matter of days, and may act with such rapidity so as to preempt any immunological defences generated by the host, must have the ability to endure decades in suspended animation, or else infect very different hosts such as cattle, in order to finally be able to infect another hapless human who is wandering by. Yet beyond this rapid killing strategy, other strains of anthrax cause mere morbidity rather than mortality in their human hosts, so that in the end, *Bacillus anthracis* – whatever its survival strategies – must reckon with the fact that humans are one of the hosts that keep it propagating. For these reasons, McNeill and his followers feel that parasites and hosts will typically learn to live more peacefully with one another, especially after many generations – so that even malaria and plague and possibly anthrax could evolve into mere nuisances rather than dreaded diseases, like seasonal colds, and not be the scourge they have been and continue to be in some parts of the world.[4]

Nonetheless, 'evolutionary adjustment between host and parasite is complicated,' acknowledges McNeill, due in part to the diversity of available hosts, the methods by which parasites infect them, and the fact that even after millennia of cohabitation, many parasites and hosts can still wreak significant destruction on each other. There is strong evidence that parasites and hosts may be cooperative and symbiotic with each other for a period, and later become competitive and antagonistic towards one another, depending on changes that may be linked to diet, climate, social organization or the behaviour of other parasites and other hosts, along with many other factors. As one group of evolutionary biologists point out, 'Species interactions have evolved from antagonistic to mutualistic and back several times throughout life's history.' Bringing such insights to the global questions of human history could mean that even if indigenous peoples of the Americas and Australia–New Zealand had somehow been periodically exposed to Old World microbes during pre-colonial centuries, it may not be true that their immune systems would have been better prepared to resist their European conquerors. The project of acquiring immunity depends on nature as well as nurture, and the interactions of both.[5]

These intricacies of the parasite's dilemma therefore complicate the assumption that parasite and host necessarily become more friendly with each other over time. The benefits that a parasite can offer a host, from killing the host's enemies to supplanting more vicious parasites to offering the host certain curative powers, means that the parasite may view the host as the principal freeloader in their relationship. In the case of Cortés and his men invading Tenochtitlán, these Spanish hosts depended heavily on their parasites, and such parasites might rightly have expected more from their hosts, even exploiting and sickening them in a sort of payment that they extracted for all the services they provided them. The implications of game theory means that individual parasites tend to push their greediness as far as it will go, since individual defectors in symbiotic relationships reap more rewards, at least temporarily, thereby enabling their progeny to become more successful over time while inflicting greater costs on future hosts. Here is the parasite's dilemma played out at the level of biochemical pathways,

DNA mutations and gene fitness. It seems that because of a parasite's greed, the world may never evolve into a disease-free place, since some parasites could never envision themselves as getting a fair deal for the services that they provide to their hosts. All of this is to say that, from the perspective of your average bodily parasite, we human hosts would seem to be the real parasite in this union.[6]

One therefore needs to consider the dilemma confronted not just by parasites, but also by hosts, since hosts need to decide how far they are willing to tolerate their parasites. Too much suffering inflicted by the malaria parasite means that people and their immune systems will seek all means possible to eliminate plasmodia from their insides, even if some tolerance can prove very beneficial, as by sickening their enemies. It also turns out that a host's effort to battle and expel a parasite can be more troublesome than simply learning to live with it: *resisting* a parasite can be more stressful than *tolerating* it. Or, as a group of parasitologists see it, a host's 'resistance and tolerance should be seen as complementary, not alternative, defense strategies'. A host's attitude towards its parasite therefore depends on how much harm the parasite inflicts, as well as how much benefit it provides. In the end, asymptomatic parasitaemia, as the state of being parasitized without experiencing any negative side effects, may be less a condition in which parasite and host give something to the other happily and willingly, and more one in which they do it begrudgingly and obligatorily. A seemingly cooperative, mutualistic relationship may be much more coercive than it appears. Or at least this mutual coercion could be key to understanding the parasite's world view. The typical parasite may have little choice but to infect a host at the same time as trying to please it.[7]

If *parà-sítos* (from ancient Greek) is anyone who eats beside, then *hospes* (from medieval Latin) is anyone involved in hospitality, whether as host or guest. Even though many languages such as English distinguish *host* from *guest*, just as German distinguishes *Gastgeber* from *Gast*, Latin makes no distinction between those giving shelter and food and those receiving it. In this way, the Italian and French *ospite* and *hôte*, as derivatives of *hospes,* are terms that denote both host and guest. The seemingly distinct roles of host and guest can be united under the concept of hospitality, or *hospitium* – as in the form of hotels, hospices and hospitals, which bring together servers as well as served, owners as well as visitors, hosts as well as guests. Philosopher Jacques Derrida exposes the dilemma of hospitality, whereby a host feels an obligation to take in a guest or foreigner, but always at the risk that the guest overstays his or her welcome and begins consuming more resources than can be offered. Here, ongoing and inevitable tensions arise between hospitality and hostility, says Derrida, 'because the one who invites may turn into a hostage of the guest' – which is to say that the host must welcome the guest but always does so begrudgingly. Or, as Michel Serres adds,

> It might be dangerous not to decide who is the host and who is the guest, who gives and who receives, who is the parasite and who is the *table d'hôte,* who has the gift and who has the loss, and where hostility begins within hospitality.[8]

A parasite is therefore choosing to live with the host, who is obliged to take the parasite in, but each of them suspects that the other is reaping greater advantage in their union.[9]

This etymological excursion reveals that parasitism has close parallels to hospitality. In the world of hospitality, the *maître d'hôtel* (or host) considers the guest as someone who has the tendency to become a parasite, just as in the world of parasitism, the parasite views the host as someone who may receive more than he or she deserves. Both host and parasite are committed to their relationship, for each derives key benefits from the other, but each regards the other suspiciously. Tension thus prevails between the members. The project of resolving the parasite's dilemma then becomes a challenging and prolonged process of deciding how much to compensate the host, since the host will suffer if the parasite is too selfish or will reap excess advantage if the parasite is too generous. An Italian adage teaches that *L'ospite é come il pesce:*

dopo tre giorni puzza – 'A host is like a fish; after three days it stinks.' Yet according to the double meaning of the term *ospite*, after three days of staying with each other, *both* host and guest become smelly to the other.

Likewise in the world of human disease, host and parasite form relationships whereby each member aims to benefit from the other, but with the union wavering between mutual trust and mutual distrust. One could say that relationships between host and parasite shift from cooperation to competition and back again. McNeill recounts how syphilis, as the western hemisphere's probable gift to the rest of the world, struck with diminishing severity after arriving on Old World shores. During its first European introduction in the sixteenth century, it was reported that syphilis victims typically suffered dramatic 'ful-minating symptoms' – but that in subsequent centuries the disease became less devastating, with *Treponema pallidum* and humans devel-oping more tolerance for one another. Initial hostility between these organisms morphed into greater tolerance. The dynamic nature of host–parasite relationships may mean that hosts and parasites take time to adjust to each other, and that this adjustment is expressed as disease, which varies according to tolerance, individual and circumstances. Or as two medical scientists counsel, 'Ditch the term pathogen . . . Instead of focusing on what microbes do or do not do, researchers should ask whether an interaction between a host and a microbe damages the host, and if so, how.' To appreciate this interaction, one must therefore think like a host, but also like a parasite.[10]

Two's Company, Three's a Crowd

A parasite's life in a host becomes more complicated but potentially more influential when a third species joins a simple twosome. In this enlarged union, a parasite may be involved with an intermediate host as well as its main host, and then seek to provide benefits to them all for coaxing all members to stay faithful in the bigger relationship. The bubonic plague of the fourteenth century killed through virulent bacteria, passed to a human by a flea carried by a rat, so that all three creatures – bacteria, flea, rat – served to propagate human disease, and actually involved four creatures if humans served as reservoirs for the pathogen. Typhus likewise stems from bacteria riding a flea (or a mite or louse), which rides a rat to a human while infecting them all in the process. Indeed, 'four's a crowd' if we envision the *Toxmoplasma*–rat–cat–human union as the team responsible for transmitting Toxmoplasmosis to people, along with possible mind-altering effects. This parasite that may be promoting greater risk-taking in men and higher aggression in women depends on at least three non-human hosts to carry out these services. A parasite may therefore be obligate to a collection of hosts to complete its life cycle and propagate its benefits. One may question whether all the hosts in these symbiotic mega-unions are really benefiting in their relationships with the parasites that they nurture and pass on. Can the rat that infects a person with the plague derive any benefit from the flea that it carries or from the bacteria that the flea carries? Does the mosquito that infects a person with malaria derive any benefit from the plasmodia that it carries? It turns out that in many cases there are indeed trickle-down benefits for parasites that are hosting parasites.[11]

One of the most devastating creatures to join parasite–host couples in human history is the mosquito. Mosquitoes serve as the third member of various three-part unions that are responsible for at least a dozen serious human diseases, from malaria and dengue fever to yellow fever, Chikungunya and Zika, all of which depend on these winged insects to transport microscopic creatures between people. Each year, the mosquito's ability to transmit microbes from infected to uninfected humans means that millions of people die, tens of millions become ill and hundreds of millions become asymptomatic hosts who feel no ill effects. McNeill describes the transmission process as one of microbes 'hitching a ride' on a mosquito, with the insect just happening to take a blood meal before flying to the next host. Likewise,

Jared Diamond in his bestselling *Guns, Germs and Steel* declares that pathogenic microbes get a 'free ride' from their mosquito carriers. But truth be told, the *Plasmodium* parasite has long made sure that its transmission by a mosquito is no haphazard process, since the *Plasmodium* is neither waiting passively for its next transmitter, nor is the mosquito offering its transportation services for free. To garner the transportation services of the mosquito, *Plasmodium* needs to offer benefits to its flying host, just as it needs to provide benefits to us, its sedentary host.[12]

It turns out that the mosquito is laying out significant resources to accommodate its demanding parasite, since *Plasmodium* busily completes its life cycle and reproduces within the mosquito's body and is therefore a drain on the insect's metabolism. The mosquito does not get very sick from its parasite – if it did, it could not fly very well to deliver its payload to the next human – but a mosquito's physiology is slightly impaired by this protozoan. *Plasmodium* 'is known to cause a decrease in fecundity and fertility of mosquitoes', according to one study. It is therefore revealing to explore how this parasite may be compensating the mosquito for its energy consumption. Interestingly, there is strong evidence that even though a *Plasmodium*-infected mosquito suffers diminished fecundity, it simultaneously enjoys greater longevity – and so will ultimately have greater ability to incubate the parasite and pass it on to the next human. *Plasmodium* can apparently manipulate the mosquito's physiology so that it will divert energy from reproduction into longer life – even if another group of scientists point out that a laboratory's enhanced growing conditions may be the main reason why these mosquitoes are living longer. But if it really is true that *Plasmodium* helps mosquitoes live longer, then we need to wonder whether the same parasite may be prolonging lifespans of the other creatures that it inhabits, notably a large, rather hairless, bipedal mammal. Even though it is unclear exactly how *Plasmodium* lengthens a mosquito's lifespan, measurements in the laboratory show that *Plasmodium*-infected mosquitoes live longer on average than if they had never encountered the parasite. Could the same mechanisms be working in *Plasmodium*-infected people?[13]

Still other intriguing evidence that mosquitoes reap important benefits from the plasmodia that parasitize them stems from observations in Kenya of two large tents that each house sleeping children, with one tent sheltering malarial individuals and other sheltering non-malarial ones. In the morning, the malarial tent has invariably attracted many more mosquitoes than the non-malarial tent, with the explanation being that the plasmodia in the malarial children's blood emit odours through human skin that, like a magnet, lure the mosquito to come and take a blood meal. Without *Plasmodium* taking up residence in the human bloodstream, the mosquito is less likely to find its human host, and less likely to be able to suck up the liquid meal that is crucial

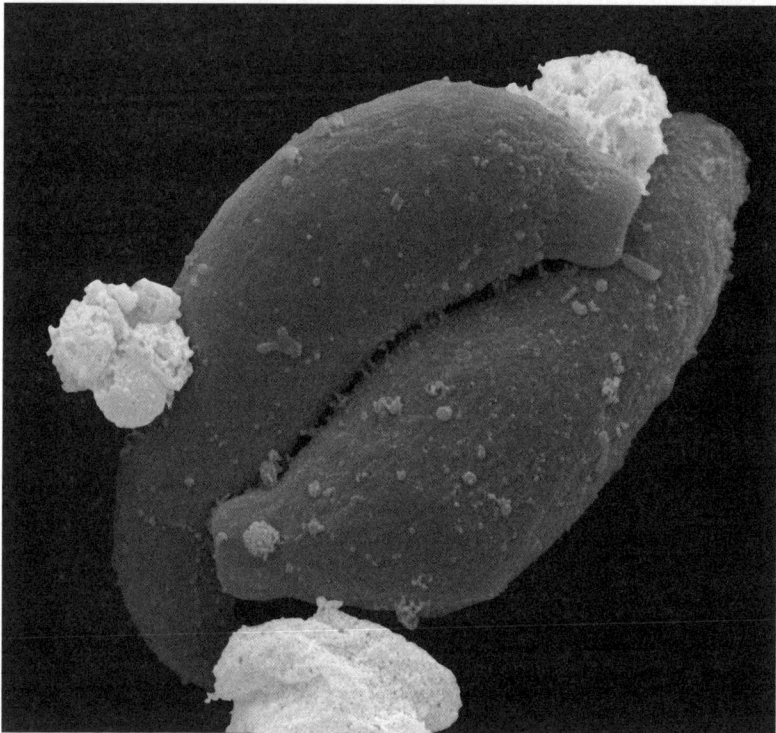

Scanning electron microscope image of two *Plasmodium* ookinetes, a developmental stage of the *Plasmodium* found within a mosquito. An ookinete is about one-tenth the width of a human hair.

to the mosquito's own reproductive success of developing and hatching eggs. So, here again, a parasitic relationship assumed to be detrimental to the mosquito host is in fact mutualistic since the *Plasmodium* helps the mosquito locate human blood, with the mosquito in turn helping the *Plasmodium* by transporting it to the next person and continue its life cycle. In this scenario, the mosquito and *Plasmodium* are benefiting each other.[14]

The ability to enhance longevity and induce attractive odours are just two ways by which *Plasmodium* is apparently paying back the mosquito for the nutrients it consumes and the transportation services it receives. Little surprise that mosquitoes may be receiving still other advantages from their *Plasmodium*, such as being able to withstand greater levels of nutrient deprivation, with researchers concluding that '*Plasmodium* infection alters metabolic pathways in mosquitoes . . . thereby conferring a survival advantage to the insects during periods of starvation.' Other studies add that *Plasmodium*-carrying mosquitoes are more likely than their non-carrying cohorts to seek out and feed on the next human host, with the little protozoan implicated in making its mosquito host more aggressive at biting, even if the 'precise mechanism responsible for increased anthropophagy in sporozoite-infected mosquitoes is not yet clear'. While modern science struggles to uncover the precise biochemical pathways by which malaria's parasite benefits malaria's transmitter, we can assume that the protozoan as well as its insect host are also cautious about offering these generous gifts to each other, always suspecting that their partner may be taking more than it gives. If the tensions of hospitality also hold true in this relationship, the mosquito is wary of becoming too debilitated by the parasite that it carries; at the same time, the parasite is concerned that the mosquito won't be able to carry its payload to the next human. Yet as long as both creatures continue receiving vital services from the other, both will continue doing their best to foster their give-and-take relationships.[15]

To turn harder into human history while remembering these biological insights, we know that a parasite like *Plasmodium* can make us very ill – but it can also benefit us by maintaining our immunities, infecting our enemies, protecting us from more dangerous microbes and even levelling the economic playing field by sickening rich and poor alike. We can now join the benefits that people receive from the *Plasmodium* with the benefits that *Plasmodium* receives from the mosquito to realize that both creatures are offering benefits to us. One can also appreciate that when the parasite evolved to carry out some of its crucial life stages within a mosquito, it was in effect sprouting wings that could carry it to the next human. *Plasmodium* had found that two hosts are better than one – with this parasite hardly hitching free rides without compensating its transporter.

Palaeoepidemiologists believe that in evolutionary time, reptiles and birds, followed by non-human primates such as chimpanzees and gorillas, were the first to host *Plasmodium* before this parasite began inhabiting human beings. Molecular evidence also indicates that certain malaria parasites, such as the *Plasmodium falciparum* – humanity's deadliest – have only been transported by mosquitoes for the comparatively short time of around 10,000 years, out of the 200,000 years that human beings have walked the Earth. Other, less virulent *Plasmodium* species have a longer experience living within their human hosts. To then enrich this evolutionary story, it may be that *Plasmodium* was initially hosted only by mosquitoes, its definitive host, without having any access to a larger animal to serve as an intermediate host. Clay Huff was one of the first researchers to suggest that these intermediate hosts may therefore have arrived last into what was initially an exclusive *Plasmodium*–mosquito affair, with both members eventually realizing that reptiles, birds and primates are lumbering bags of blood that are made even more lethargic and easier to target by a well-placed parasite – with the result that a *Plasmodium* and a mosquito joined up for exploiting these large-animal hosts. Humans, from their perspective, willingly complied with the demands of this twosome for the advantages that could be accrued, such as acquiring immunities to diseases that could then be unleashed on competitors and enemies.[16]

As recounted earlier, there is little wonder that a few Nigerians now celebrate the mosquito, viewing it as a patriotic symbol and protector of their homeland from European incursions, serving to turn their land – and much of tropical Africa – into a 'white man's grave'. U.S. Americans might also thank the mosquito for serving as a crucial ally in their war of independence, since their British adversaries suffered disproportionate vulnerabilities to mosquito-borne diseases, with malaria leading the charge. Sardinians, too, were able to maintain a stronghold on their island over the centuries through the assistance of their malaria-carrying mosquitoes that fended off invaders trying to make beachheads. In 1323, during a siege of Cagliari by the Aragons of Barcelona, 'an epidemic exploded across the army that was so serious as to make everyone sick,' noted Pietro IV, the future king of Aragon. 'The epidemic lasted as long as the siege . . . And in this way, it can be calculated that half of the attackers died during the siege, leaving very few of the other half free from the fevers.' Although the Aragons eventually prevailed to take Sardinia's main city, the diseases they contracted made their conquest difficult and deadly. One can now appreciate that in all these human battles, parasite-carrying mosquitoes, by providing their more familiar hosts with better immunities, had their own motives for choosing sides in these human conflicts: the people who fed and carried their parasites for the longest were endowed with greatest measures of defence.[17]

One can now understand why mosquitoes can themselves be considered parasites, and not just hosts: mosquitoes are hosts to *Plasmodium*, and parasites to people when they suck blood and feed on a person's nutrients. It follows that humans contracting malaria can view themselves as being accosted by *two* parasites, one protozoal and the other arthropodal, with people supplying foodstuff to both and becoming ill in the process. 'Host' and 'parasite' are therefore ambiguous labels for all members of this *affaire à trois*: each organism sees the other two as the real beneficiaries, as the true parasites. In this larger stage of hospitality that involves a union of not two but three *hospes*,

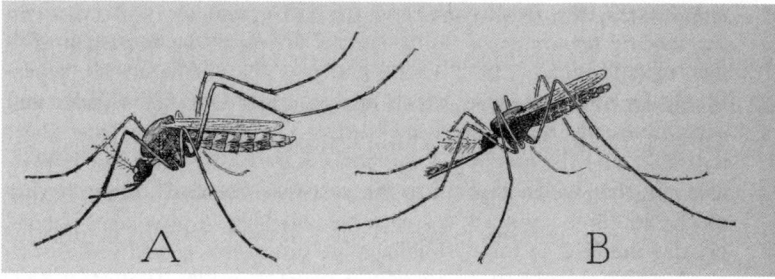

Two mosquito species: *Aedes aegypti* (A), which carries *Flavivirus* of yellow fever, and *Anopheles punctipennis* (B), which carries *Plasmodium vivax* of malaria.

the mosquito, *Plasmodium* and human are each playing the roles of both host and guest. In the struggle to stay alive, reproduce and live out our ambitions, the act of joining up with compatible partners is a crucial strategy for us all. One might say that mosquitoes are a social species, since they need their own kind as well as their symbionts in the form of *Plasmodium* and people. Likewise, we humans benefit from the company of our own species together with our various symbionts, since they make life for us easier and even possible. Every second of every day, groups of species, small and large, depend on each other in joint projects of conviviality. As a group of community ecologists recently asserted, their field 'remains narrowly focused on antagonistic interactions among species ... [when in fact] facilitatitve interactions are common and strong among coinfecting symbionts'. In Donna Haraway's terminology, humans, mosquitoes and plasmodia can be considered 'companion species'. Or, as science historian Gregg Mitman emphasizes, 'the biological is social all the way down.' In the end, the labels of 'parasite' or 'host' may say more about an organism's relative size than about the role each plays in a relationship, since each of them depend on the other. To a parasite, a host seems parasitic, even if it is very large.[18]

Across deep time, a select group of mosquitoes have proven to be some of our most faithful fellow travellers, often moving with us wherever we move. Only two hundred out of the 3,500 mosquito species actually feed on people. As we migrated across continents and across the oceans, these mosquitoes have often buzzed along behind us, seeing in us the source of their next blood meal. In fact, it has been a vital decision by these mosquitoes to feed preferentially on humans, generally avoiding the tempting blood-filled cow or cat or rat that also travels alongside us. Even if there are various species of malaria parasites that can infect sundry lizards, birds and mammals, from gibbons to bats, the mosquitoes that chose to hitch themselves to our star have been extraordinarily successful, spreading to every continent, even displacing native mosquitoes that were already living there. This propensity to move with people means that many of humanity's most concerning mosquito-borne diseases are not transmitted by native mosquitoes, but by those alien to many of our homelands. *Aedes albopictus*, the Asian tiger mosquito, is native to central Asia and has followed people across the globe to transmit dengue and Chikungunya viruses in newly colonized places. Meanwhile, *Aedes aegypti*, the yellow fever mosquito, has spread out of Africa to colonize wide swathes of the Americas and bring the fever there. *Anopheles gambiae*, also native to Africa, is a principal vector of malaria that was spread by people around the world.[19]

But influencing the feeding preferences of humanity's key co-travellers has been no haphazard process, since humans have themselves encouraged these little bloodsuckers to come along for the ride: people removed dense forests, developed water-dependent agriculture and constructed dwellings to deflect rainwater, concentrating it into puddles and storing it in receptacles, thereby enlarging mosquito habitat on many fronts, and so spreading the parasites that these mosquitoes transmit. Human bodies have also served as water and energy sources in arid places, so that when humans moved there, so could the mosquitoes that required these resources. This caravan of people-plus-mosquitoes-plus-microbes travelled to the far corners of the Earth, as along the coasts of southern Africa, where these trios had previously been scarce. *Aedes aegypti*, for example, took advantage of human-stored water 'where doing so provided the only means to survive the long, hot dry season'. The very project of agriculture, by turning over soils, meant that rain more easily collected in small puddles of exposed dirt rather than being absorbed by unbroken soil. In more recent times, large construction projects, such as the building of the Panama Canal, relied on heavy equipment and high-pressure sprays to excavate foundations and heap up dirt, serving to multiply the standing water that is so crucial to mosquito larvae. In 1912, an entomologist stationed on the isthmus noted that the malaria-carrying mosquito, *Anopheles albimanus*, 'is closely associated with man and finds its most congenial surroundings about his habitations and in conditions he creates in the course of agricultural, engineering and other work'. Looking towards the future, it has been noted that 'rapid urbanization currently taking place in Africa will drive further mosquito evolution, causing a shift toward human-biting in many large cities by 2050.'[20]

Perhaps more faithful than our household dogs and barnyard cats, our human-associated mosquitoes are some of our most intimate domestic pets, always remaining obedient – even tame – as we moved around the Earth. From our own selfish perspective, we created the habitats for our trusted mosquitoes for the benefits they can provide us, such as fending off subsequent invaders, with our domesticated mosquitoes being more vicious than our guard dogs. A biologist defines domestication as 'a coevolutionary process that arises from a mutualism, in which one species constructs an environment where it actively manages both the survival and reproduction of another species in order to provide the former with resources and/or services'. That our closest mosquitoes fit this description so nicely suggests that we may as well fit them with little collars, while giving them cuddly names such as 'Buzzy' and 'Spotty'. Our faithful mosquitoes will remain allied to us as long as we continue providing benefits to them – and we will probably continue feeding and housing them for as long as they keep providing benefits to us.[21]

Medical entomologists typically label *Aedes aegypti* as *anthropophilic* – human-loving. But the lengths to which we humans have gone

Humans have long created ideal habitat for many mosquito species, as seen in this woodcut from *Ortus sanitatis* (1491).

to construct ideal habitats for these creatures and feed them would suggest that the mosquito is the one in charge, commanding us to build it a luxurious home. Barring a few odd insecticide sprayers and pond drainers, the fact that people are so quick to comply with the desires of our mosquito friends could mean that we are *arthropophilic* to them – insect loving. If humans experience a 'botany of desire', as Michael Pollan teaches, whereby our useful and desirable plants are convincing us to cultivate them, then mosquitoes are promoting a 'zoology of desire', since they have been so skilful at convincing us to build them

178

habitats and expose our veins to them – even as they sicken and kill some of us in their larger project of assisting the rest of us. Haraway once declared that 'if I have a dog, my dog has a human.' By corollary, it seems that if I attract a mosquito, then my mosquito attracts me.[22]

Historian Urmi Engineer Willoughby has traced colonial settlement patterns in the Caribbean and the southern United States to find that wherever people stopped long enough to construct ports and build homesteads, *A. aegypti* also travelled to these areas to call these lands home. Since *A. aegypti* is the primary vector of yellow fever, Willoughby has relied on historic outbreaks of this disease to track where this mosquito moved across this region, finding that it journeyed 'from Havana to Port-au-Prince, from Charleston to Boston':

> By the end of the eighteenth century, colonial port cities up and down North America's Atlantic seaboard supported large populations of *Ae. aegypti*, which were becoming naturalized in their new urban habitat ... Ongoing immigrant waves of non-immune peoples created ongoing conditions for epidemics: human blood provided mosquitoes with nutrition and city architecture provided them with habitat ... The spread of urban *Ae. aegypti* populations after the American Civil War led to the nation's worst yellow fever epidemic, which after breaking out in 1878 radiated up from New Orleans throughout the Mississippi Valley, causing pain and suffering as far north as Chicago and Pittsburgh.[23]

But disease-carrying mosquitoes did not inflict serious harm on all of their human hosts who were busy expanding mosquito habitat. Greater ills were suffered by their hosts' most immuno-naive competitors and enemies, with their parasites continuing to offer their more faithful hosts a much greater degree of disease resistance.

From a socioeconomic perspective, these anthropophilic mosquitoes may have even helped disrupt macroparasitism's entrenched institutions of wealth to allow future generations of common folk to better share in society's bounties. As the 1862 yellow fever epidemic spread across Wilmington, North Carolina, it seems that the newly arrived mosquito did not single out its victims according to class. Historian Chris Fonvielle recounts:

> The disease began innocuously enough with a fever, chills and diarrhea, but eventually gave way to internal hemorrhaging, black vomit and jaundice. It did not discriminate as to its victims. Mary Ashe McRee, the wife of Dr. James F. McRee, died of yellow fever. Griffith J. McRee, a prominent citizen, lost his mother, wife, young daughter, and teenaged son and namesake. It also claimed the life of Georgia Weeks, who worked in a house of ill-repute called the Hole in the Wall. James Quigley, the superintendent of Oakdale Cemetery, succumbed on Oct. 15, 1862, his 32nd birthday. A number of local doctors, ministers and nurses risked their lives to care for and comfort the sick, and in some cases fell victim to the disease themselves. Rector Robert B. Drane of St. James Episcopal Church, Reverend John L. Pritchard of First Baptist Church, and Dr. James Dickson all died of mosquito bites while tending to patients of the so-called 'yellow jack.'[24]

Today, a different anthropophilic mosquito is arriving in the American southeast to bring the local people other kinds of ills, but also other benefits. Entomologists there report that the Asian tiger mosquito (*A. albopictus*) is now displacing the yellow fever mosquito (*A. aegypti*), with the former species much less efficient at transmitting yellow fever even if does transmit dengue fever, a less serious disease. One could therefore say that these people, having long reaped the protective benefits of *A. aegypti*, are now opening their doors to a new mosquito inhabitant who inflicts them with a less serious mosquito-borne disease. One of the great mysteries of yellow fever is that it has

Cemetery plaque commemorating victims of the 1862 yellow fever epidemic in Wilmington, North Carolina. *Aedes aegypti* appears to have been an equal-opportunity infector.

always been rare in Asia, with the best explanation being the strong presence there of the Asian tiger mosquito that tends to displace the yellow fever carrier. Moreover, the interbreeding of these two mosquito species compromises the fertility of each, with 'hybridization of the two species in zones where they overlap, producing sterile offspring'. So just as one might welcome the arrival of *Plasmodium vivax*, since it can displace or attenuate the effects of the more virulent *P. falciparum*, so too *A. albopictus* can help combat yellow fever, and even render itself sterile, thereby satisfying the goals of today's genetic engineers who are seeking to control mosquito-borne diseases by releasing artificially sterilized mosquitoes. Americans are now negotiating dengue fever but are compensated by avoiding the threat of yellow fever. Such multi-species tales demonstrate that vector biology is a complicated affair in which tweaking one end of a mosquito's ecological web will send unexpected impulses to the other end. At the very least, such stories teach that our mosquito neighbours are not always bad to all of us all the time – and that some of them generously repay the hands that feed them.[25]

If we then consider the dramatic, twentieth-century Sardinian Project, when tank-loads of DDT were sprayed across a large and rugged island to eliminate the local mosquito population, aimed specifically at eradicating the malaria vector of *Anopheles labranchiae*, we can realize why mosquitoes arriving there today are finding plenty of available habitat. It turns out that the Asian tiger mosquito is also coming fast to this island to join the abundant common mosquito, *Culex pipens*, along with a few other species, including much-diminished populations of *Anopheles labranchiae*. Thus islanders may soon be facing threats of dengue fever, even Chikungunya and Zika, which, however terrible, are less menacing than the pre-war scourge of malaria. It may be that the Rockefeller Foundation's longest-lasting accomplishment in Sardinia will be substituting a more dangerous mosquito with a less dangerous one – but one that will also be less ferocious at fending off future invaders.[26]

Parasites within Parasites

It is useful at this point to shrink ourselves still smaller, and question whether parasites like those of malaria may themselves be being parasitized. In the fashion of Russian matryoshka dolls, if people are parasitized by mosquitoes and mosquitoes are parasitized by plasmodia, then one rightly suspects that still smaller creatures are living in or on *Plasmodium* – and possibly even smaller entities may be living in those creatures. It would then follow that these mini-parasites would also need to be careful not to offend their hosts, and indeed, could not act in a truly parasitical way, at least for very long, or their relationships would break down. The rules of hospitality would hold true for these smaller pairs of *hospes*.

Here we enter the cutting edge of molecular biology and scanning electron microscopes to consider the world of viruses, organelles and plasmids. Viruses are organized packets of DNA or RNA that are surrounded by protective protein coats; organelles are functional subcellular units containing genetic material; plasmids are simple strands of nucleic acid (DNA or RNA). All of these subcellular particles might be considered parasitic structures, since they cannot survive without the help of their larger host cells, and need to anneal themselves to a prokaryote's or eukaryote's reproductive machinery to generate more copies of themselves. Such cells experience something akin to asymptomatic parasitaemia if they absorb protein coats and nucleic acids and experience negligible negative effects. Bacteriophages, as bacterial viruses, can enter bacteria, possibly disrupting them and causing them to lyse, or break apart, but sometimes providing them with extra abilities to increase their fitness. Thus bacteriophages alternate (according to circumstance) between being competitors and cooperators with their bacterial hosts. Bacteriophages can also serve to reinforce their larger human host's health, as by enhancing the resistance of bacteria they infect, which are responsible for aiding crucial functions such as digestion – meaning they can be mutualists to both bacteria and humans.[27]

Certain bacteriophages are labelled by microbiologists as 'domesticated' if they promote their bacterial host's abilities to reap more resources. Just as people have domesticated wolves into faithful dogs by providing them with a home and food, so too bacteria have domesticated rogue bacteriophages into docile forms that assist their host to surmount its own challenges. Here again are a parasite and host morphing over time into two mutualists that assist one another. A current aspiration of bacteriologists is to synthesize bacteriophages that can reinstate antibiotic sensitivity in unwanted bacteria, with humans in effect aiming to disrupt that mutualist relationship for maintaining the power of antibiotics.[28]

In organisms larger than bacteria, such as eukaryotes in the form of *Plasmodium*, there are subcellular plasmids and organelles that regulate the cell's physiology, all of them reproducing as the eukaryote itself reproduces. Because they assist their mother cell, these plasmids and organelles are mutualistic rather than parasitic, having come into their roles through extended cohabitation, or symbiosis, which favoured reciprocal benefit.

If one then views these subcellular, gene-carrying organelles as 'selfish genes' in the tradition of Richard Dawkins, then one is endorsing a deterministic process of DNA choosing to reproduce more successfully than another strand of DNA. The project of 'thinking like a parasite' may therefore find its barest expression as a functional sequence of nucleotides – or gene – that is ultimately and mechanically concerned only with itself. Yet it is precisely through such bare egotism that a selfish gene is concerned about the organisms it depends upon, otherwise it could not survive and thrive. Even at the molecular level, self-interest breeds altruism. A selfish gene promotes its host cell, which in turn can promote a multicellular organism, and with it the species, and indeed the larger community. As Dawkins put it, 'mutual benefit will evolve if each partner can get more out than he puts in.' One can therefore feel very good about a selfish universe – as long as this universe is co-dependent – since every creature that depends on another creature aims to make its partner live happily ever after. If we believe that we are surrounded only by selfish genes, then, all the way down to our nucleotides, we really are a cooperative lot.[29]

Although some believe in the existence of a *cooperative* gene (or 'greenbeard' gene in Dawkins's terminology), others counter that there is, at best, 'preliminary evidence' for this kind of gene. It makes easier conceptual sense to envision cooperation at the genetic level as stemming from mutually dependent genes, with each relying on self-interest to ensure that its partner benefits in the relationship. So, too, at the organismal level, a mutually dependent parasite and host aim to make sure that the other reaps benefits so that their relationship will continue.

Or to turn to the scenario of ancient Greece, a (human) parasite offered witty conversation to its host in exchange for a delicious feast, with both parasite and host happy to continue their arrangement for as long as the other kept giving. It may be that *Plasmodium* and *Homo sapiens* will also be happy to continue their arrangement, despite all the pain, since *Plasmodium* is no simple human pathogen.[30]

Even if 'apicoplast' is not your everyday household term, there is a single one of these minuscule structures sitting inside every *Plasmodium* cell, busily carrying out vital functions for its host. In classic symbiotic fashion, the apicoplast is an organelle that receives nutriment and shelter from its eukaryotic host. It then returns these favours by manufacturing and distributing such biochemical wonders as isoprenoids, fatty acids, iron-sulphur clusters and heme. As an endosymbiont, the apicoplast is thought to have once lived independently in its primordial state, surviving by its own energies through photosynthesis. Eventually it reaped the benefits of entering and living permanently within the *Plasmodium* while adapting its own functions to assist its host, even reproducing itself while accompanying the *Plasmodium* through its complicated developmental stages that progress through a mosquito to a human and back again. Such is the magic of biology, which some mosquito manipulators want to harness for crippling this endosymbiont, disabling the parasite and eliminating malaria. The apicoplast keeps the *Plasmodium* satisfied, and so researchers see a method for disease control by disrupting it. But there is still the usual challenge of targeting a single element within an intricately interdependent biological system, whereby damage inflicted on the apicoplast potentially spreads beyond its host to damage its host's host (the mosquito), and even its host's host's host – people. In meddling with apicoplasts, molecular biologists should be reminded of John Muir's dictum that when we try to pick out anything by itself, we will find that it is attached to everything else in the universe.[31]

Yet another key endosymbiont that has captured the attention of mosquito-borne disease experts is *Wolbachia*, a bacterium that is found in up to 70 per cent of all insect species, including key mosquito species. These bacteria have the amazing property of being able to neutralize many human pathogens carried by the mosquito, especially viral pathogens. Thus *Anopheles* mosquitoes can carry *Plasmodium* as well as *Wolbachia* along with hundreds of other eukaryotic symbionts, some 453, according to one count. Indeed, just as the human body plays host to a cornucopia of creatures living within it, every mosquito carries untold numbers of creatures within its small frame, with scientists now actively cataloguing this insect's virome, bacteriome, fungome and eukaryome.[32]

Complicating this symbiotic plot is the fact that one of humanity's most troublesome mosquitoes, *Aedes aegypti*, does not ordinarily carry the *Wolbachia* bacteria. However, it can be induced to do so, and thus help guard against the human transmission of several viral diseases. A series of clever discoveries by mosquito biologists has shown that lacing *A. aegypti* eggs with *Wolbachia* will transmit these bacteria to the maturing mosquito, and so serve to denature the problematic viruses it transmits – which are pathogenic to people as well as to the mosquito: *A. aegypti* infected with Chikungunya virus, for example, experiences important 'fitness costs' and lays fewer eggs. Simply release a batch of *A. aegypti* infected with *Wolbachia*, and Chikungunya – along with dengue and, to a lesser extent, yellow fever and Zika – begin to fade away. With an apparent mutualistic concern for its mosquito host, *Wolbachia* is able to neutralize the pathogenic viruses, thereby protecting mosquito and people alike. Surely here is an endosymbiont working hard to protect both its definitive and intermediate hosts. Little surprise, then, that *Wolbachia*-infected *A. aegypti* are now being released around the world by the boxload. Not a dog, not an anthropophilic mosquito, but a bacterial endosymbiont, may be man's best friend.[33]

The triumphalist tone of medical entomologists who point to the success of *Wolbachia* may be tempered by realizing that controlling disease by propagating endosymbionts is really a project of biocontrol, which entails its own risks. When an exogenous (or alien) organism is released into an area to fight another organism, there is a chance that this novel creature will run out of control – rather like how cane toads were released in northeastern Australia in 1935 to control sugarcane pests, with the ecologically disastrous result that the toads multiplied exponentially to overrun much of that continent. Unforeseen social and political consequences, such as distrust in agricultural authorities, accompanied these ecological disruptions. Any endosymbiont thinking (and acting) like a parasite could, without significant foresight and planning, end up thinking like an alien species, and sally forth to multiply and inhabit the bodies not just of other mosquitoes but of other insects that don't ordinarily play host to these bacteria – ultimately creating dramatic imbalances in natural and cultural systems. Various risk assessments have been compiled in anticipation of releasing *Wolbachia*-laden mosquitoes around the world, with the conclusion being that there are probably few dangers, though ongoing monitoring is mandatory. A multi-expert panel from Indonesia recently declared that the release of *Wolbachia*-laden *A. aegypti* in their country would entail 'negligible risk' even if 'a strong and rigorous Monitoring Plan should be an integral part of the project.' *Wolbachia* offers much promise in controlling mosquito-borne diseases, but this bacterium may not yet be the magic bullet its boosters claim it to be.[34]

Ultimately one must marvel at the image of a world filled with endosymbionts, which biologist Lynn Margulis celebrated as a fundamental source of biological evolution. Symbiogenesis is considered by Margulis and her followers as the process by which smaller creatures move inside larger creatures, with the organisms attracted to these unions by the benefits they can both accrue – while providing quantum leaps in biodiversity. Two billion years ago the most complicated life forms were the relatively simple prokaryotic cells of bacteria and archaea, but various novel symbiogenetic events then created more complicated cells in the form of eukaryotes. Multicellular eukaryotes followed, and the rest is history – with the diversification of plants and animals producing the modern tree of life. What may have been motivated by the selfish, parasitic behaviour of one creature as it moved into and reaped the resources of another eventually evolved into symbiotic, mutualistic behaviour after the newly arriving organism realized that harming its host ultimately harmed itself. In this evolutionary progression, parasitism leads to mutualism, despite occasional defectors that threaten a stable union. Ongoing natural selection assures that true parasites are continually being eliminated in a give-and-take process that favours cooperators over defectors. In the case of *Wolbachia*, this bacterium is thought to have once been a parasite of the mosquito, and then became its mutualist, at least under most conditions. Although these bacteria can be 'notorious for their reproductive parasitism' of insects, says one group of researchers, they also 'engage in mutualistic relationships with their hosts'. Cooperation eventually trumps competition, it seems, but it can be a rocky road to this harmonious state – which may then be disrupted by the next environmental shock.[35]

The theory of symbiogenesis therefore supports the view that through evolutionary time, not only did plasmids move into viruses, and viruses (as bacteriophages) into bacteria, but bacteria moved into eukaryotes (or *proto*eukaryotes) to become indispensable organelles such as chloroplasts and mitochondria, which carry out photosynthesis and generate energy in these larger and more complicated organisms. Although it is still unclear how these different-sized partners joined one another, whether through initial cohabitation or immediate engulfment by phagocytosis, the partners became single organismal units. Such eukaryotic unions would themselves join together to create multicellular organisms, some cells of which would specialize and carry out vital functions for their larger community of cells. Social theorist Peter Kropotkin would surely have welcomed the views of Margulis and her followers, since he saw all organisms, people included, as necessarily joining together in mutual assistance to help one another and build stable systems and communities. From this line of reasoning, mutualism is manifested as cooperation at all levels of life, from organelles to republics, so that there can be no such thing as parasites – at least over the long term. It seems that every creature and every person will always be working to satisfy the needs of its host, always being ready to serve them. Not only have we long served our malaria parasites; they have long been serving us.[36]

Symbiogenesis happens not only in the natural world, but increasingly in the laboratory, for molecular engineers are now inserting genetic elements into various kinds of cells, with the goal of creating viable and useful new symbionts such as the *Wolbachia*-infested mosquito, along with other surgically spliced creatures. Although genetically modified organisms (GMOs) can now be found on every grocery shelf and in nearly every farmer's barnyard and field, there is a strong interest in creating new parasite–host unions within cells. Through increasingly sophisticated processes called 'artificial endosymbiosis', scientists are pushing strands of DNA (in the form of plasmids), and whole viruses (in the form of a bacteriophage) into bacterial genomes, hoping to manufacture desirable creatures coded and produced by these new insertions. They are also injecting whole bacteria into larger cells for the aim of synthesizing new organelles in what are sometimes called 'designer endosymbionts'. By creating, in effect, artificial parasites and artificial hosts, the scientists are hoping that these partners will be able to form stable, mutualistic unions. This quest to synthesize novel endosymbionts, explains one study, requires attention to syntrophy as well as tolerance: 'Syntrophy requires that neither partner can survive without the other ... Tolerance refers to the symbionts having either negligible or manageable adverse effects on the survival and fitness of their symbiotic counterparts.' Such wording signals that the scientists are really recreating the details of hospitality, whereby host and parasite can welcome each other, with both of them learning to tolerate the other through the benefits that each accrues. Symbiogenetic engineers are learning to think like a parasite.[37]

Although the project of 'directed symbiosis' is certainly more challenging to carry out in multicellular than bacterial systems, there is much excitement at the prospect of forging brand-new endosymbionts, since such creatures might be harnessed to produce new kinds or greater quantities of food, biofuels, fertilizers, pharmaceuticals, bioremediants and much more. Not just in animal systems, say the boosters, but in the plant world, these new symbioses will reap enormous benefits. There is already evidence, for example, that synthetic chloroplasts fashioned out of cyanobacteria can be inserted into larger cells to allow them to carry out photosynthesis, thereby reproducing events like those that apparently happened more than a billion years ago when cells incorporated functional chloroplasts into their tissues. As one study announces, 'molecular biologists, biochemists, bioinformaticians, system biologists, and synthetic biologists' will all be needed to carry out and assess this work – to which we can suggest adding economists, political scientists, ethicists, philosophers, historians and even theologians – if we really begin seeing artificial parasites and artificial hosts flowing from our labs. The ramifications of such creatures on natural and human systems could be momentous but also irreversible. These new endosymbioses will require close monitoring, since they are meant to satisfy one key *exo*-symbiont in the form of us – people – who are supposed to benefit from such manipulations. The potential for dangerous endosymbiotic side-effects means that we had better think twice before playing God with parasitism.[38]

Multiparasitic Ethnographies

In Sardinia, they eat wormy cheese. It is a delicacy so teeming with crawling creatures that European Union regulators won't even allow it to be sold to the public, even if scattered farmers across the island continue to produce the stuff in backyard bins, always ready to pull out chunks of it on special occasions to offer to relatives and close friends. The cheesemaker typically wields a broad knife to cut away the crust, exposing a whitish-grey paste with little channels that on closer inspection house pulsating, twitching worms – which are more visible in some sections than others. If the worms do not twitch, the conscientious cheesemaker will withdraw her offering, for only live worms mean that this is high-quality 'casu marzu' – *formaggio marcio* – or rotten cheese. Refrigerated *casu marzu* typically contains lots of worms, all dead and resting in place, which may be relished anyway if fresh rotten cheese is unavailable. But connoisseurs wait until the waning summer months when *Piophila casei*, the tiny cheese fly, has been given free rein to insert its eggs just under the surface of seasoned blocks of curdled sheep milk. These then hatch into larvae, or maggots, which proceed to burrow throughout the ageing cheese, simultaneously digesting it while excreting intricate trails of tangy paste. At this point in the cheesemaking process, one can scoop up the paste, worms and all, to be slathered over a slice of country bread. Maybe here at last we have found the secret to Sardinia's amazing longevity, for fresh *casu marzu* can be found nowhere else.[39]

There are other wormy cheeses produced across Italy, and wherever else cultured milk is left to season further in the open air, but Sardinia's version is perhaps the most famous, with the *Guinness World Records* once labelling it 'the most dangerous cheese to human health'. But locals who consume the wriggling paste and offer samples to adventurous tasters claim that 'no one has ever been made ill by eating it' – and indeed counter that only good comes out of consuming generous proportions of what their parents and grandparents also relied on to spice up their own meals. Certainly, a laboratory analysis of Sardinia's human microbiome would produce a list of the bacteria and protozoa that habitually inhabit this morass of rotting cheese, which would include the creatures that live inside the cheese fly, as well as its maggots and all of their own parasites. It turns out that this fly is a fairly cosmopolitan creature, commonly feeding on carrion and other meat products to the consternation of butchers around the world. Forensic entomologists even rely on the lifespan of the fly's larvae to calculate the age of cadavers. The fly and its pulsating maggots might not be considered parasites in the tradition of a tapeworm that spends its whole life within a well-stocked intestine of a horse or human, yet, since these flies depend heavily on human food products, they may be considered macroparasites adapted to human environments, especially farmsteads littered with milk and meat products. The gift that the fly gives back to its Sardinian host for the calories it consumes is of course a bountiful harvest of rotten cheese. Both fly and human hold up their ends of the deal in this mutualistic union. To extend this multispecies ethnography, the fly and human are joined through the cheese and its *Lactobacillus* to the sheep and the grass they grazed on, and also to the farmer and tourist and possibly to the mosquito and its malaria parasites, to highlight a few of the many creatures that compose this pastoral ecosystem. These creatures all depend on each other in symbiotic networks, with parasites and hosts, micro and macro, long benefiting from each other through the rotting cheese of Sardinia. Anna Tsing would point out that 'interspecies entanglements that once seemed the stuff of fables are now materials for serious discussion.'[40]

No wonder that historian Carlo Ginzburg focused so much attention on wormy cheese in his microhistory of northern Italy, *The Cheese and the Worms*, which examines the beliefs of Menocchio, a mystical sixteenth-century peasant from a small mountain village. In Menocchio's world view, ageing cheese spontaneously gives rise to worms, which

Casu marzu, a cured sheep cheese further cured with cheese fly maggots.

Shepherd preparing sheep cheese in Sardinia, 1950.

appear from one day to the next to take on a life of their own. The worms become eternal, even deistic, through the powers that they exhibit:

> earth, air, water, and fire were mixed together; and out of that bulk a mass formed – just as cheese is made out of milk – and worms appeared in it, and these were the angels. The most holy majesty decreed that these should be God and the angels, and among that number of angels, there was also God, he too having been created out of that mass at the same time, and he was made lord . . .

These exuberant declarations of a medieval mind, which challenged the day's ecclesiastic authority to earn the charge of heresy, venerated the squirming little creatures and the relationships that they embody. The opportunistic cheese fly's larva, burrowing through a farmer's curdled milk, is a precious member of Earth's community, even holy in its powers to appear out of nowhere. Here, the worm is no pest or enemy but ally and friend, even saviour, whether tunnelling through a block of rancid cheese or attached to the insides of a goat's or a person's intestine. These testimonies from five centuries ago celebrate a wormy parasite because it contributes vitality to the greater good.[41]

Worms, flies and beetles were found not only in rotting cheeses, but in hog pens and chicken coops, kitchens and latrines. If some farmers linked these creatures to the gods, others associated them with the diseases of their domestic animals. Tapeworms and roundworms slithering inside their pigs, cows, sheep and goats might be sapping the energy of these animals – except that these worms could also be found in the stools of their healthiest livestock. Historian John Farley points out that two centuries ago, farmers paid little attention to the insights of helminthologists, since 'certain pathologies were associated with the presence of these parasitic worms, but the worms were generally understood to be symptoms not the cause of disease.' With the advent of better magnifying glasses, the many wriggling copepods, protists and worms found living on or in living bodies began receiving greater blame for a range of maladies, animal and human. Germ theories of disease certainly contributed to parasites' ill repute by tracing so many diseases to microscopic creatures. Edoardo Perroncito, for one, followed in this tradition during the 1880s by linking the millimetre-long hookworm (*Ancylostoma duodenale*) to the terrible affliction suffered by Italian workmen who were boring the Gotthard tunnel through the Alps; he also found that extract of male fern could be useful for controlling these wormy infections even if it ended up killing some of the people who consumed it. The juvenile worm living in the tunnel's tepid and unsanitary puddles would latch on to the ankles of miners to burrow under the skin and inflict anaemia and possibly death, with one of Perroncito's autopsies revealing 1,500 hookworms lining a victim's intestines. Parasites of domestic animals were also coming to be seen as sources of human disease, even if there was no obvious mechanism for the afflictions. The fact that death and decay were associated with worms did not help their reputation. Dog historian Chris Pearson recounts how a British veterinarian at the time assumed that even if stray dogs were not rabid, their 'parasites are a nuisance, and a source of waste and insalubrity'. By the early twentieth century, medical and agricultural parasitologists saw their main mission as killing worms and any other slimy creatures that lived in or on people and their domesticated friends. Health required 'sanitation', which meant applying antiseptics, cleansers and disinfectants – and any other substance that might be injected, ingested or rubbed on for killing bodily critters.[42]

But with the rise of mass pest control through chemicals, a few observers deemed certain parasites to be more 'successful' than others, since some of them were more numerous, lived longer or apparently inflicted less damage on their hosts. In 1922 an American science writer offered that 'no successful parasite can annihilate its hosts, but it can materially impair their health and reduce the numbers of their individuals.' A passage from a Melbourne agricultural journal clarified that 'The successful parasite exploits its host only to a degree that the exploitation can be borne. It is wary of "killing the goose that lays the golden egg"'. One of the day's best-known sympathizers of parasites was certainly Theobald Smith, the microbiologist who led the charge that parasites, despite their labels, were not always despicable freeloaders that caused disease. For Smith, parasites typically fostered a 'certain equilibrium' with their hosts and only when pushed off this balance did disease occur. Smith felt that parasites inflicted progressively less injury on their hosts over time, with any troubling side-effects of their union eventually fading away. But most other parasitologists displayed their self-centredness by envisioning parasitism only from the perspective of the host, interpreting disease as a parasite inflicting pain rather than a complicated relationship of hospitality.[43]

As medical historian Pierre-Olivier Méthot recounts, the Law of Declining Virulence rose to its heyday in the twentieth century's interwar period. It began to attract criticism when parasitologists and pathologists were presented with field and laboratory evidence that pathogenic organisms did not always lose their virulence, at least not regularly and consistently. Ongoing experience showed that some parasitic diseases persisted, returned erratically or even displayed greater virulence over time, as was the case for certain episodes of malaria or yellow fever. As microbiologist Hans Zinsser pointed out, 'parasitic adaptations are not static, and . . . extraordinarily slight changes in mutual adjustment between parasite and host may profoundly alter clinical and epidemiological manifestations.' By the 1950s more and more scientists were coming to agree with him, including René Dubos, who taught that healthy humans generally develop resistance to most viruses and bacteria surrounding them, but that illness occurs 'when, for some reason, the natural resistance no longer balances the virulence of the microbes'. Mathematicians began testing this declining-virulence law, demonstrating in their formulas that host–parasite interactions might produce significant oscillations in the severity of illness. The famous case of the myxoma virus that was released to control Australia's runaway alien rabbit population offered evidence that a virus could develop into pockets of extremely toxic strains rather than into benign strains, as had been predicted by parasitologists clutching onto their assumptions about declining virulence.[44]

So, despite guarded hope that dangerous parasites and pathogens were everywhere evolving into docile mutualists, to become as subservient to their hosts as their hosts were to them, most experts had come to believe that the world would always be dealing with infectious disease. They felt that even long-running peaceful relationships between parasites and hosts would occasionally flare up to inflict serious illness. Few parasitologists kept the faith that a 'parasite's dilemma' would prevent a parasite from going out of its way to inflict severe damage on its host. And even fewer of these experts would suspect that a parasite's occasionally cruel and unpredictable virulence might be the main benefit that it could provide its host, since this virulence could be turned on enemies for reaping enormous compensation. In the next chapter, we turn to the largest host of all to see if or how its principal parasite has been able to offer it any compensation.

Rethinking the Columbian Exchange

In his bestseller *Guns, Germs, and Steel*, Jared Diamond puts microbial pathogens at the forefront of his explanations for why Europe was able to race ahead of other regions and subjugate distant peoples, seize their lands, join continents in trade and build modern economies, finally amassing a disproportionate share of political power and material wealth to dominate the world. While guns and steel stemming from fortuitous inventions and plentiful resources, along with geography, certainly aided the Western juggernaut, argues Diamond, most of Europe's (and to some extent, Eurasia's) expansion was made possible because Europeans were habitually exposed to dangerous viruses, bacteria and protozoans, and, after building significant resistance to these microbes, unleashed them on the rest of the world. Where Alfred Crosby famously coined the term 'Columbian Exchange', which concept he would develop as a drawn-out process of ecological imperialism that relied on living creatures, large and small, that spread out to Europeanize much of the world's natural and cultural systems, Diamond clung to the explanatory power of germs. Where Crosby noted that 'ecological stability did tend to create a crude kind of mutual toleration between human host and parasite,' Diamond considered deadly epidemics as a leading factor 'Underlying the Broadest Pattern of History'.[45]

In Diamond's reliance on the pathogenic role of microbes in his grand theories, he offers little suggestion that parasites and even pathogens may have had silver linings, able to bring benefit and not just harm.

With his conviction that a germ is just 'some damned clever microbe [that] modifies our bodies or our behavior such that we become enlisted to spread microbes', Diamond sees little complexity in a microbe's role, since a person's main goal of staying healthy is 'best done by killing the damned germs'. Although Diamond does acknowledge the parasite's dilemma by asking, 'Why should a germ evolve the apparently self-defeating strategy of killing its host?', his answer is that the microbe's deadliness is merely an 'unintended by-product', meaning that he does not recognize that the vast majority of microbes aim to benefit their hosts, directly or indirectly, over the short or long term, and generally make a habit of trying to keep their host happy. Diamond never came to terms with the fact that a host's bodily creatures will disappear if their host disappears.[46]

If we can begin to think like a parasite, and recognize some of the many ways by which parasites benefit their hosts, we can appreciate that the little organisms living in our gut, liver or bloodstream see themselves as performing important services for us, such as threatening our adversaries, promoting our acquired and innate defences against diseases that others remain susceptible to, or producing high fevers that kill off some of our more problematic microbes. Since not everyone is equally susceptible to the same microbe, the more resilient among us are given important advantages, such as the Amazon's Yanomami people, whose resistance to measles has been deemed as strong as that of most Europeans. Beyond these benefits we reap from our pathogenic parasites, we can also acknowledge the advantages provided by our mutualistic parasites, which can displace dangerous parasites or even mitigate destructive anti-immune responses stemming from excessive harm inflicted on our normal bodily fauna. Other creatures living within us are only occasionally problematic, such as *Bacteroides thetaiotaomicron*, which, while sporadically irritating our stomach walls, also synthesizes vital vitamin B12. Although some parasites certainly maim or kill their hosts, most of them are also quite blind to race, class or gender, and inflict their ills indiscriminately to help level society's unjust playing fields. Only with a paradigm shift – acknowledging that parasites bring benefit and not just harm, promote longer life and not just quicker death – can we understand the larger role of 'germs' in human affairs beyond that of simple emissaries of death.[47]

In the Columbian exchanges of the past five hundred years, Eurasian microbes did not just inflict pain and suffering on non-Eurasian societies to usher in our modern age of Western hegemony. Many of our parasites and even our pathogens were not as dangerous as they appeared, turning out to serve both colonizer and colonized as mutualist organisms that aided people wherever they lived, if not all of the time, then most of the time. In fact, many of Jared Diamond's 'germs' turn out not to be germs at all, at least in an exclusively pathogenic sense. By making the deadliness of microbes a primary driver of the human experience, Diamond falls far short in offering a convincing global history. William McNeill once cautioned that 'Diamond's reduction of the tangled web of recorded history . . . is a clever caricature rather than a serious effort to understand what happened across the centuries and millennia of world history.' David S. Jones added that one 'must acknowledge the ways in which multiple factors, especially social forces and human agency, shaped the epidemics of encounter and colonization'. To these criticisms, one can add that many of the microbes that people exchanged across continents and oceans were vital to both donors and recipients. And while certain disease-causing parasites and pathogens were deadly in their results, such microbes were not acting out of malice towards a foreign host but out of concern for a familiar one.[48]

6

Parasites R Us

There is little possible doubt that the overall impact of the human
species on the planet is quite literally (and not just metaphorically)
parasitical.

KENT PEACOCK, 'Sustainability as Symbiosis', *Alternatives*, 21 (1995)

seudomonas syringae may be the most misunderstood bacterium on the planet. Ever since the early 1900s, when blights on
wheat, clover, onion, tomatoes, fruit trees and many other
crops were traced to this rod-shaped bacterium, farmers have been
searching for ways to eradicate it or at least remove it as best they can
from their fields: they have drilled pesticides into soils, sprayed copper
solutions onto leaves and coated seeds with antibiotics. But, knowing
what we now know about parasitic microbes, it should not be surprising to find out that *P. syringae* does not always cause harm to its plant
hosts, and, according to a variety of observations, may in fact be providing them with hidden benefits – as by promoting plant growth or
else providing biocontrol of various plant pathogens. However it
cannot be denied that one of *P. syringae*'s most irksome properties is
its unusual ability to facilitate freezing in plant extremities, so that in
any normal crop harbouring these bacteria, a brief cold snap during
the growing season can result in brutally frozen leaves, burst plant cells
and failed crops – with *P. syringae* then multiplying over these ice-
damaged tissues to cause more damage. And yet, even *P. syringae*'s
ability to heighten a plant's susceptibility to frost may not just be for

its own benefit, but for the benefit of nearby organisms and indeed of the whole Earth, us included.[1]

David Sands, an agricultural scientist working in Montana, was reportedly the first to come upon the idea that *P. syringae* may be promoting more good than bad. In 2012, some three decades after his original revelation about the possible silver lining of this feigned plant pathogen, Sands described in a lively TED talk how he came upon the solution to the mystery of the reappearing *P. syringae* in a farmer's wheat field even though all precautions had been taken to eliminate it from the seeds and soils before planting time. After receiving a troubling phone call from a farmer who reported that *Pseudomonas* – aka 'Sue' – had surely reinfested his fields, Sands and his team travelled up from Bozeman to try to find out how this bacteria had found its way back into the crop. In searching for the source of the bacterial bloom, Sands explained that after some deliberation they looked left and they looked right and finally looked up at the sky, to decide that what they really needed was a plane:

> So we flew up there holding the petri dish out the window. Let me tell you, in a Cessna 180, holding the petri dish out the window, and changing it every 500 feet, and circling over the field and trying to avoid the vomit bag. It's all very cool. And so we did this. And, we found the bacteria, not in the air, but in the clouds; and not just in the clouds but in the ice crystals. And there it was. And so we developed this idea: that the bacteria live on plants, and they got to get off those plants and go somewhere else . . . They make a million bacteria overnight. OK? And then they decide to leave and fly up into the air, just like balloons. They are very, very light. And they go wherever the wind takes them. Aha! So now we have a hunch, which I will call a hypothesis . . . Hmm, it's the bioprecipitation cycle: bacteria live on plants; often they don't hurt the plants; they are swept into the air; they cause ice crystals in the clouds; the

ice bounces around; and down it comes as rain or snow. And that's pretty cool. We liked that. Nobody else did.[2]

Sands was too modest, since he and his colleagues were finally able to publish their theory of bioprecipitation in a 1982 issue of *Időjárás: Az Országos Meteorológiai Szolgálat folyóirata* (Journal of the Hungarian Meteorological Service). It was here that the Montana scientists announced to the world that 'The implications of a two-way relationship between rain and bacteria are far-reaching.' It would take more than two decades for the rest of the scientific community to begin accepting their proposal that *P. syringae* is actually more mutualistic than pathogenic, and that climate itself is being modified by microbial organisms. A definitive study would corroborate that '*P. syringae* plays an active role in the weather cycle' by inducing precipitation onto the very plants that play host to these bacteria. So here was another piece of evidence that life on Earth modifies global processes to favour life on Earth. James Lovelock called these biophysical feedback processes the Gaia Theory. Lynn Margulis called Earth a symbiotic planet.[3]

'Ice nucleation' is the label for the process of water vapour crystalizing around cloud-borne particles, such as bacteria, dust or pollen, and is a crucial step in the water cycle whereby plants nurture microscopic creatures before releasing them skyward for seeding clouds to precipitate moisture back upon themselves and their surroundings. Although ice nucleation has been understood since the 1960s, when clouds were artificially seeded with silver iodide particles from aeroplanes to produce rain and snow, only recently have airborne bacteria been appreciated for their role in inducing precipitation in the same way. Like a cow that nurtures gut microbes that help digest the hay it swallows, a grassy field hosts *P. syringae* before sending them skyward to precipitate moisture and help supply itself with water. So effective is this bacterium at precipitating water and ice that ski resorts add it (and extracts of its cell walls) directly to the water flowing through their snow-making machines. Indeed, a cottage industry of sorts has cropped up in our era of global warming for spraying artificial snow over ski runs, with the whole process depending heavily on *P. syringae's* ability to align water molecules on its cell walls to favour ice crystallization. A group of meteorologists who have analysed snow samples from around the globe claims that bacterial ice nucleation 'is widely dispersed in the atmosphere' and provides crucial 'feedbacks between the biosphere and climate'.[4]

And so, if a Montana wheat field can host bacteria to modify local weather and enhance growing conditions, then there are surely many other organisms from marine algae and rainforests to arthropods and mammals that are also altering ecosystems for their own benefit – especially by hosting symbionts that mediate these benefits. We humans certainly modify our own niches to make life easier on ourselves, building homes, developing agriculture and harnessing fossil fuels, along with all the other ways that our day-to-day activities refashion the natural world to support our modern lifestyles. Humans have become environmental engineers on a grand scale, terraforming the planet to make it more habitable for ourselves and the creatures that we depend upon. So transformed is Earth by human action that many believe we have entered a global human age, or Anthropocene, that can be traced to the artefacts of civilization that are being preserved in the geologic strata. Palaeontologists of the future will locate our own global imprint in the sediments and soils beneath our feet.

But is it really possible that we humans and our microbial co-travellers have made Earth more favourable for life, including our own? Or is humanity taking more than giving, corrupting more than improving, becoming a kind of macroparasite on the earthly system? As Michel Serres believes, 'history hides the fact that man is the universal parasite, that everything and everyone around him is a hospitable space. Plants and animals are always his hosts; man is always necessarily their guest.' In the following pages I explore the implications of humans as parasites, not in their interactions with other people, but in their biophysical role of interacting with the whole planet. If it seems that our species takes more than it receives, then we also need to wonder if there are unexpected or hidden ways that humanity is also giving back to the global system so as to compensate its macrohost. As is the case for so many parasitic organisms that we have encountered in previous chapters, humanity may be providing hidden benefits to its host, making Gaia more productive, more inhabitable and possibly healthier for itself and other forms of life. Projecting parasitism from microcosm to macrocosm requires a leap of faith, but we need to wonder whether relationships of hosts and their dependents may also hold true at very large scales.[5]

'Rain Follows the Plough'

As North America's westward-bound settlers marched across the Great Plains in the nineteenth century, they encountered increasingly arid conditions quite unlike those they were used to in the continent's well-watered eastern seaboard and Europe. So, too, Australian settlers pressing into the Outback witnessed aridity unlike anything they knew on their eastern shores. In fact, marginally habitable deserts and dry-lands around the world have been common destinations for some of the latest waves of human settlement, from the New World's interiors and Africa's arid lands to the Middle East and northeastern Asia. In all these areas, boosters and government bureaus have often been quick to dispel the hardships of building new lives in these desolate places. Contrary to what seem the simple hard truths of unproductive desert life, these promoters encouraged newcomers to keep coming, bring their livestock, build houses, till the soil and plant their crops. Nature will do the rest, they said: 'Rain Follows the Plough'.

Historian Henry Nash Smith has traced how this 'meteorological fantasy', as he put it, became ingrained in these settlers' beliefs. Not only profiteers and politicians but many respected scientists had come to assume that human activities and the habits of agricultural life, especially the planting of crops and forests, would serve to attract

moisture-laden clouds to unleash life-giving rain. An 1880 edition of
Scientific American quipped that because 'rain follows the plow, as it has
before', new farms had been successfully established across America's
arid prairies in a belt 350 miles wide. In responding to an early twentieth-
century questionnaire issued by the Public Land Commission about
possible climatic changes, more than a quarter of the Nebraskans sur-
veyed agreed that their summers had become wetter since they arrived
and began tilling and planting the fields a few decades earlier. In noting
that the western desert's land-locked Great Salt Lake had actually risen
a metre or so since Mormon pioneers had begun settling the area a generation earlier, such a respected investigator as John Wesley Powell felt
that the best explanation stemmed from 'human activities such as farming, cattle-raising, and cutting trees'. Not until the 1930s did this 'folk
belief' begin to die, noted Smith, especially as more and more of these
settlers witnessed their farms dry up in a new era of drought.[6]

Yet folk beliefs may bear kernels of truth. As bioprecipitation theorists now argue, significant moisture really does fall from clouds
seeded with *Pseudomonas syringae* and other ice-nucleating bacteria
emanating from underlying crops. Spontaneous and cultivated vegetation attracts at least some of its own rain and snow. Depending also on
local topography, humidity, wind and water bodies, discrete amounts
of moisture crystallized by airborne microbes fall on the plants that
host them, with the rest falling further afield. Researchers focusing on
central Asia conclude that some of the area's recent irrigation projects
have 'significantly increased local precipitation'. Another study of
North America's Great Plains claims that areas recently converted to
agriculture have witnessed midsummer rainfall increase by 15 to 30 per
cent. Defoliation and deforestation often have the reverse effect, with
the clearing of vegetation apparently leading to desiccation and drought,
even if there are exceptions, as at Colorado's Wagon Wheel Gap experiment, where denuding forests a century ago reportedly increased total
water runoff. Although any one location's total precipitation and its
runoff depends on much more than airborne, ice-nucleating bacteria,

there is little question that humans, plants and animals, together with their microbes, all interact with the surrounding soils, winds and sunlight to modify global processes. Such observations mean that the biotic and abiotic are modifying each other: biology and geology are linked sciences.[7]

Lovelock and Margulis developed a vision of Gaia around such ideas, whereby Earth's organic and inorganic elements form a dynamic feedback system that helps create and maintain favourable living conditions. Earth scientist Lovelock and microbiologist Margulis pointed out that photosynthesizers generate oxygen, carbohydrates and nutrients that are then cycled back into useful forms for all of Earth's creatures. Sun energy drives moisture into clouds, and bioprecipitation then directs part of this moisture onto the plants and animals that need it. Across cosmic time, biological processes have produced high atmospheric levels of oxygen and nitrogen, steady concentrations of salt in the oceans and relatively favourable physiological temperatures. People, acting as both consumers and producers, have become part of this homeostatic system, by means such as promoting rainfall through farming.

Yet both Lovelock and Margulis were initially less certain about whether or not modern humans could significantly modify the larger processes of Gaia. At first they viewed humanity as just another top-level omnivore having rather minor effects on the functioning of global ecosystems, largely unable to either disrupt or accelerate primary physical and biological processes. Writing in the mid-1980s, about a decade after the unveiling of the Gaia hypothesis, Lovelock considered the evidence to be 'very weak indeed' that 'industrial activities either at their present level or in the immediate future may endanger the life of Gaia as a whole'. Margulis concurred that, as humans, 'We cannot put an end to nature; we can only pose a threat to ourselves.' Yet over the next decades, both scientists would come to believe that *Homo sapiens* was pushing dynamic equilibria beyond their Gaian limits. Accumulating levels of air and water pollution, expanding islands of discarded plastic,

James Lovelock and Lynn Margulis under the gaze of goddess Gaia, 1983.

accelerating soil erosion and species extinctions and greenhouse-gas emissions were all signs that our own species was no longer a neutral biotic force, much less an advantageous mutualist that contributed to the smooth functioning of a symbiotic planet. Lovelock came to declare that 'we are unintentionally at war with Gaia,' with Margulis agreeing that 'Western culture is expert at the abuse of our planet's water heritage.' By the twenty-first century, both scientists were emphasizing that our species was lurching beyond any role of promoting a healthy Gaia to emerge as its most self-serving member, one that was ruthlessly taking more from the earthly system than it was giving. One need only glance around to realize that humanity had ruptured its benign relationship with its Gaian host, becoming the dictionary definition of a parasite: *a creature that lives on or in another species and derives its nutrition and/or protection, often with harmful effects to the host.* By endangering its host, humanity was ultimately endangering itself – finally confronting the hard truths of the parasite's dilemma. We have so 'grievously damaged the Earth', warned Lovelock, 'that Gaia now threatens us with the ultimate punishment of extinction'.[8]

A Living Planet?

The possibility that humanity may be acting parasitically on a global system would imply that our planet is a living being, or at least may really be alive. The earthly host, if acting as an organism, could theoretically be born, grow, react, heal and even die, with other organisms able to harm or consume it or even cooperate with it. Yet an orbiting and revolving planetary body enveloped by a largely nitrogen-oxygen atmosphere overlying water-filled basins would seem capable of harbouring life, but not be life itself. Margulis worked hard to counter Lovelock's early tendency to envision Gaia as an enormous living being. In the close working relationship that the two developed over the decades, this was one of their main disagreements, with Margulis emphasizing, 'I have always clearly maintained that "the Earth is a single live organism" is *not* the Gaia idea.' She taught that the earthly system relies on dynamic feedback loops of which living creatures are a part, but she never considered Gaia (as some of her critics felt she did) to be a sort of cosmic being floating in space. She celebrated the deistic metaphor of Gaia for the traction it garnered in the public mind, but she never saw the thing that it described to be a celestial organism with a will of its own. Margulis was much more comfortable with a symbiotic vision of Gaia, whereby networks of creatures, or 'holobionts' as she called them, worked together to promote a self-sustaining synergy. She was also keenly interested in the origins of life, which seemed to rely on many of the same processes that would favour a global steady state and might lead to inhabitation and eventually to relationships of parasitism and mutualism. 'What is life?' she once asked. 'It is a material process, sifting and surfing over matter like a strange, slow wave.'[9]

Lovelock loved to tell the story of how a casual conversation with his literary neighbour, William Golding, had served as the inspiration for naming his theory after the Greek goddess Gaia to encapsulate the notion of Earth as a mythological deity. Yet Lovelock would spend much of the rest of his career reshaping the contours of this life form after realizing that many of his peers hesitated to take any Gaia Hypothesis seriously, since it implied a spiritual rather than a positivist view of the global system. He sometimes countered that if Gaia is not organismal, it is certainly life-*like*, or at least that 'the Earth might in certain ways be alive,' adding that it was our own limited understanding of life that prevented us from recognizing Gaia's version of it. As a physical scientist, he praised the ideas of Erwin Schrödinger, who taught that life in its most basic form is something that counters entropy, or the tendency of things to spontaneously dissipate and lose energy. Living entities distinguish themselves, taught Schrödinger, by feeding upon the 'negative entropy' of systems in higher energy states. In comparing the atmospheres of our nearest planets with that of Earth, Lovelock theorized that Earth's high levels of gaseous nitrogen and oxygen could not have arisen through inert chemical reactions alone but depended on Earth's photosynthesizing creatures to harness sunlight and convert inorganic ingredients into vital organic compounds and byproducts. Lovelock added that these biological and physical reactions served to accumulate substances in the land, sea and air that were favourable to life, which had not been the case in such non-Gaian planets as Mars and Venus. Agreeing with proponents of the bioprecipitation theory, Lovelock envisioned the role of Gaia's global mantle of forests as both absorbing and generating rainfall: 'Without the trees there is no rain, and without the rain there are no trees.'[10]

In their search to identify the type of living entity that is Gaia, Lovelock and Margulis lauded Vladimir Vernadsky's notion of a 'biosphere', which he developed in the 1920s from the term coined by Eduard Suess a generation earlier. Vernadsky, a Russian Ukrainian polymathic scientist, considered the biosphere to be that portion of Earth formed by living processes, such that 'Two distinct types of matter, inert and living... exert a reciprocal action upon one another.' Expanding on Suess's description of the biosphere as a sort of living film wrapping Earth, Vernadsky emphasized that organismal activities joined with geological processes for producing Earth's subsurface, surface and atmosphere. Even though Lovelock confided that Vernadsky's ideas were somewhat speculative when they first appeared, he felt that he and Margulis had 'retrod much of the path [Vernadsky] laid', having 'discovered him to be our most illustrious predecessor'.[11]

Microbial mats, as curious geological structures originating from living processes, were a sort of Vernadsky Object of special interest to both scientists. Such mats are not simply transformed remains of diatoms or corals, as is the case for limestones, for example. Nor are these mats deposits of fossil fragments, since they are neither original organismal parts nor replicas of such parts replaced by minerals. Such mats are instead the product of chemical reactions generated through the activities of cyanobacteria and archaea interacting with various substrates, and have been going on since the Precambrian – or the earliest stages of Earth's formation. Famous examples of microbial mats include so-called stromatolites, which are bulbous outgrowths of rocks, isolated or extensive, that have long mystified geologists and are typically found in shallow marine environments along with a few freshwater ones. Perhaps it is stromatolites, then, as a kind of growing rock, that provide evidence for Gaia being alive. A Gaian planet may not exhibit all the traits of typical living organisms, but it embodies some traits of life while displaying others special to itself.[12]

To emphasize that Earth's physical and biological components interact with each other, Lovelock and Margulis began portraying the science of Gaia in medical terminology, employing such words as 'geophysiology' or 'planetary physiology'. Here was an investigative field first articulated by geologist James Hutton that sought to understand how inanimate substrates stemmed from biological processes. From the perspective of Gaia theorists, Earth physiologists (like their plant or animal physiologist counterparts) aim to understand the functions and activities of the whole planet, while identifying key tissues, organs and systems that rely on chemical reactions and feedback loops. Biological signals are sent across terrestrial spaces, with recipient structures communicating with donor structures to maintain a healthily

Stromatolites found at Shark Bay, Western Australia.

running network. Lovelock outlined his musings on this subject in *Healing Gaia* (1991), where he explored how the earthly system might adapt to or recover from the permutations exerted by its most demanding member, *Homo sapiens*. His recommended antidote was to favour strategies that would rein in the excesses of this rogue species, or at least curtail its most disruptive activities. Lovelock acknowledged that waste is a normal byproduct of living organisms, for example, but that humans were producing such enormous and poisonous quantities of it that the whole Gaian system was suffering. Margulis rejoined that humans, like all life forms, cannot expand limitlessly: 'Nor can we continue to destroy the other beings upon whom we ultimately depend.' She emphasized that our species relies on co-dependency between all life forms, from the smallest subcellular organisms to the whole biosphere.[13]

On Stewarding the Earth

So if one can entertain the notion that the earthly system called Gaia, if not alive, at least exhibits living traits, then it becomes more reasonable to suggest that humanity, now surpassing 8 billion people, plays a role in the functioning of Gaia, be it positive or negative, ameliorating or damaging, mutualistic or parasitic. If Gaia is more than a metaphor for describing a dynamic homeostatic global system, then *Homo sapiens* in the twenty-first century can be considered more than just a simple component of this system, being a species that may be creating an indelible geological epoch named after itself. A geophysiologist investigating the functioning of Earth today cannot ignore the activities of people.

Paul Crutzen, the Nobel atmospheric chemist and one of the first to propose that our Earth has entered a geological human age, cultivated a long working relationship with James Lovelock. As early as 1974, Crutzen was alerting his senior colleague to his 'interest in the way in which agricultural processes may affect the atmosphere', thereby reflecting the rising attention to ideas about bioprecipitation and their relations to an early Gaia Hypothesis. Crutzen was still seeking out Lovelock's opinions a decade later; for example, over the question of whether nuclear warfare might load so many particulates into the atmosphere as to 'reduce crop growth and biospheric productivity'. Even if Lovelock in these years was sceptical that any human activity could have lasting global effects, he nonetheless agreed with 'the spirit and intention' of Crutzen's claims. Yet, in considering the corollary that human activities might even be harnessed for improving rather than corrupting Gaia's functions, Lovelock insisted that *Homo sapiens* would make a terrible Earth steward. Nearing the end of his 103 years, and finally admitting that human activities posed a severe threat to a healthily functioning Gaia, Lovelock felt that humanity would never be able to properly manage Earth: 'I would sooner expect a goat to succeed as a gardener than expect humans to become stewards of the Earth.' If humanity hoped to engineer a troubled Gaia, it would have, as Lovelock titled his last full book, *A Rough Ride to the Future*. Margulis likewise taught that 'The planet takes care of us, not we of it.'[14]

Stewart Brand, another Lovelock correspondent and friend, famously declared, 'We are as Gods, and may as well get good at it.' But despite the wild popularity of Brand's *Whole Earth Catalogue* (1968–98), Lovelock was quite convinced that humans playing God would never get very good at taking over and improving upon global processes. Lecturing on 'The fallible concept of Stewardship of the Earth', Lovelock emphasized the enormous challenges in managing the natural world while professing much more faith in Gaia's own workings. He warned his audiences about relying on 'breathtaking hubris as the technological fix':

> Too much CO_2 in the air? – send out the tankers full of iron chloride to feed the algae that will then remove the CO_2. This is the environmental equivalent of gunboat diplomacy. Occasionally it might be the right thing to do but more often, it is the harbinger of a gaudy and costly mess.[15]

Lovelock saw more hope in 'planetary medicine' than stewardship. Since we cannot begin to understand all of Gaia's workings, we must aim to address humanity's most egregious activities, such as by limiting our worst effluents and relying on empirical approaches to identifying what we need to do. Gaia will remain in charge, noted Lovelock, with the best strategy being to step back and allow natural processes to run their course.

Dividing Gaia into so-called *strong* and *weak* forms was proposed by those who aimed to distinguish the Earth system's degree of intentionality. A strong Gaia (or at least a stronger Gaia) presumed that the global system could deliberately and purposely activate processes and events that favoured its own needs, as opposed to a weak Gaia that operated through passive and spontaneous processes. Weak Gaia could be better tested and investigated, and was easier to accept for most scientists, whereas strong Gaia indicated greater abilities to adapt and show vitality – indeed, be more *lifelike*. Weak Gaians reasoned that the geosphere and biosphere adjusted to each other according to homeostatic feedback mechanisms, whereas strong Gaians felt that Earth's biotic and abiotic processes could optimize conditions for life on Earth. Strong Gaians more easily envisioned Earth as a superorganism, and so considered humanity as a crucial member of a system that could manipulate and even manage its own global environment. As climate scientist Daniel Lashof explained, biogeochemical feedbacks are 'mediated by one of the most interesting organisms on the planet – *Homo sapiens*. Climate change affects human activity in a myriad of ways, and thus influences anthropogenic emissions of greenhouse gases.' In Lashof's view, people might spoil their own nest as well as make progress towards cleaning it up.[16]

A deliberate, strong Gaia thus gives more agency to human abilities to manage Earth, which also encompassed Vernadsky's concept of the Noosphere, or that part of the Earth system arising out of human reason and consciousness. Building on Pierre Teilhard de Chardin's vision of the noosphere as a global manifestation of human ideas,

A Gaia theorist's view of Earth, generated with AI: 'Gaia unveiled – vivid watercolour depiction of Earth as a thriving organism with interconnected ecosystems'.

Roman depiction of Gaia, the Greek goddess of Earth, 1st century CE.

Vernadsky offered that the act of cutting down forests, for example, would

> change the face of the planet, create numberless new physical-chemical processes in the history of the biosphere, until now acting more or less unconsciously. In the noosphere, the regulating of this function of humans must be one of the basic features of its new structure.[17]

In other words, human thoughts and ideas were integral to a self-regulating and self-sustaining Earth system. In the final years of his life, even Lovelock came to believe that Gaia might benefit from a sort of global network of intelligence: 'We alone, among all the species that have benefitted from the flood of energy from the Sun, are the ones who evolved with the ability to transmute the flood of photons into bits of information gathered in a way that empowers evolution.' It seems that humanity, perhaps never very good at stewarding Earth, might nonetheless improve its abilities at doing so through the wonders of supercomputing and artificial intelligence.[18]

Gaia Engineering

A strong Gaia that incorporates our human species as key members therefore ushers in a new era of global engineering whereby humans can hone Earth into a more inhabitable system. Earth engineers not only seek to mediate global change, but aim to make good decisions about those changes. But as Lovelock points out, hubris is the knee-jerk judgement of many our bio-geoengineering schemes. Ever since Archimedes boasted at being able to move Earth if he could only fix a stable fulcrum, men (mostly) and their machines have been proposing any number of Jules Verne scenarios for anchoring sand dunes, irrigating deserts and rearranging seas. François Roudaire's 1870s proposals for excavating canals for draining Mediterranean sea water into vast basins lying below sea level in North Africa relied on large quantities of hubris. In this way, Roudaire aimed to create inland seas that would moderate Algeria's and Tunisia's climates, while opening arid lands to agriculture and settlement. Yet amazingly, similar schemes actually succeeded. Canals were built across Denmark and the Suez and Panama; whole lakes were drained in Italy and central Asia and California; dams were erected across the Nile, Mississippi and Yangtze. The Soviet Union's ambassador to the United Nations in 1949, Andrei Vyshinsky, justified his country's development of nuclear weapons by noting their potential for 'blowing up mountains, changing the course of rivers, irrigating deserts, charting new paths of life in regions untrodden by human foot'. And the fact that such mega projects were merely proposed does not mean they will never be carried out. Mikhail Budyko was an early advocate of the increasingly popular proposal to spray sulphuric acid into the stratosphere for mimicking volcanic eruptions to reflect solar radiation and cool the planet. In 2023 the United States government earmarked \$1.2 billion for constructing an array of direct carbon capture devices under the assumption that human beings can actually fine-tune the atmosphere's chemical composition. Large-scale bio-geoengineering is happening all around us.[19]

From a biological perspective, humans have long been constructing their own niches. A corn field might be watered by building an irrigation canal, by seeding clouds with silver iodide, and possibly now by removing atmospheric carbon dioxide for cooling the climate and condensing water vapour. By engineering the Earth, humans have sought to make their environments more conducive to their own needs in the same way that a beaver builds a dam to form a pond for enlarging its watery habitat. As Richard Lewontin explained the niche theory, 'environments of organisms are made by the organisms themselves as a consequence of their own life activities.' It is therefore easy to believe

Artist's rendering of a Direct Air Capture CO_2 removal system.

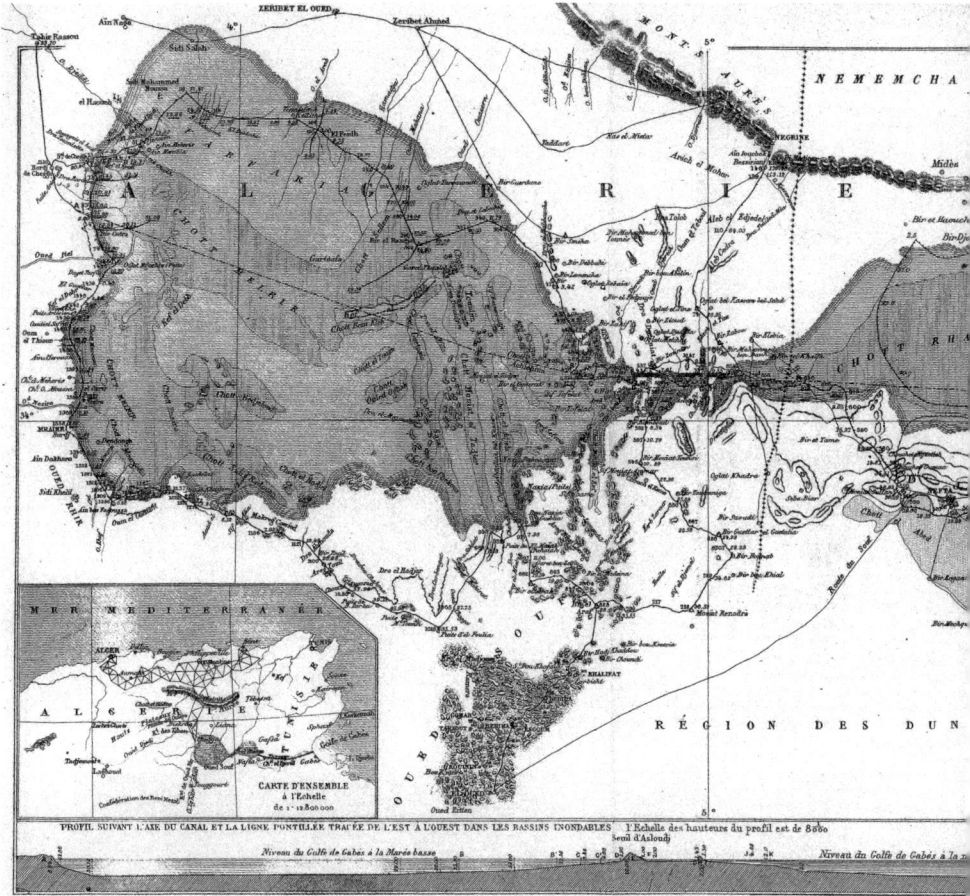

that humans are like that beaver that is busy geoforming the country-side while performing myriad agricultural and industrial activities in the project of making more food, better clothing and sturdier shelter. When people enlist tools, domestic animals and fossil fuels to assist them in these tasks, human niche construction is multiplied until vast areas of Earth's terrestrial and marine habitats have been rebuilt for human purpose. 'When our forebears became farmers they set them-selves on a path, which they are still beating out,' noted Lovelock upon reflecting on the agricultural revolution; humans 'must have had an impact on the rest of Gaia almost as revolutionary as that of the evo-lution of photosynthetic organisms millennia before'. A group of

Map of François Roudaire's 1870s proposal to drain sea water into the Chott basins of Algeria and Tunisia. The resulting inland sea would have stretched across hundreds of kilometres of the Sahara Desert.

geographers agree, noting that 'Human societies have been shaping and sustaining diverse cultural natures across most of the terrestrial biosphere for more than 12,000 years.' Viewed from deep space, Earth is a spinning blue marble, initially gassed with methane and inhabited by methanogenic archaea, then oxygenated by blue-green algae that were eventually joined by a steady march of respiring organisms, one of which morphed into a naked primate. 'Organisms do not adapt to their environments,' Lewontin asserted; 'they construct them out of the bits and pieces of the external world.'[20]

To then shrink down to a micro-organismal scale, we can realize that malaria parasites have also been constructing their world, making their immediate environments more inhabitable for themselves while also being sensitive to the needs of their hosts. Through evolutionary mechanisms – both 'weak' and 'strong' – *Plasmodium* has been able to construct molecular compounds in human veins that emit odours through our skin that can attract a nearby hungry mosquito to serve as the vector for whisking itself off to the next human being. The same *Plasmodium* is likewise constructing other niches that promote greater fitness for its human host, such as malarial resistance that is not enjoyed by the host's competitors and enemies who have never been exposed to malaria. Niche construction for a malaria parasite therefore includes rearranging human strands of DNA, with its bioengineering schemes also producing genes that code for sickle cell traits and G6PD deficiencies, for example. Humans living in malarial environments who do not carry these genetic manipulations are more susceptible to malaria's symptoms, and so cannot hope to compete against those who do. A malaria parasite aims to benefit both of its hosts, mosquitoes and people, through its intimate bioengineering of their nucleic acids.[21]

Meanwhile, *Plasmodium* is also constructing larger-scale niches through the activities set in motion by people and mosquitoes. Everything that these intermediate and definitive hosts do to the world in accommodating their parasite can be viewed as extensions of *Plasmodium*: call them 'extended phenotypes'. This parasite will modify the macro environment in myriad ways, such as by harnessing the powers of mosquitoes (for example, by inflicting disease and then inducing disease resistance) to power the humans (by conquering their enemies who lack such resistance) to power the machines that may build drainage ditches or spray pesticides or construct villages far from wetlands and other mosquito habitat. Thus a malaria parasite's extended phenotypes linked to mosquitoes and humans may serve to limit the parasite's own successes, which can be a prudent strategy for ensuring that its own hosts survive to keep itself alive.

The *Plasmodium* landscapes that have been built across such intensely malarial areas as Sardinia are the result of people aiming to avoid too many mosquitoes that carry too many malaria parasites. This little corner of Gaia has thus been heavily engineered by a parasite using its hosts as extended tools – such that the bulldozers building canals to drain watery habitat or spray insecticides are really being operated by little protozoans sitting at the controls who work to keep their own numbers limited. Even the scattered water pockets and puddles that people inadvertently create around their dwellings, which serve as mosquito habitat, and so favour mosquito parasites and thus malaria, are part of Gaia engineering. People fashion the Earth for their own needs, but these human beings are also connected to causative chains attached to more primary movers. Viewed at a molecular scale, all these malaria phenotypes can in turn be considered extensions of the genes of the *Plasmodium* that put so many entities into motion – with genes directing organelles, cells directing parasites, and hosts directing and fashioning the Earth. By honing signals, manipulating behaviours and linking social interactions, this hierarchy of extended phenotypes hones the biosphere. Richard Dawkins's selfish gene – call it a 'strong' Gaia gene if it manifests will – may well be engineering our world and, in the vision of Lovelock and Margulis, making it more fit for itself. If it turns out that humanity is acting as a parasite in a global system, then humanity remains attached to deeper nucleic acids that have long ago understood that hosts need to reap some advantage when being para-sitized, else the system breaks down. And if it is true that people today are no longer providing benefits to their earthly host, maybe the blame lies in our genes, which are becoming too selfish.[22]

Planetary Symbiosis

One can now appreciate that humans through their symbionts, through their genes, have been engineering Gaia for a long time. Archaeologists Bruce Smith and Melinda Zeder join a growing consensus that Earth's

Human Age began with the agricultural revolution. They point out that since this time around 10,000 years ago, Earth's organismal composition and physical state have been so progressively altered by our own species that one can hardly speak of any global system that does not acknowledge the human imprint. Recent climatic changes, such as those of the Little Ice Age of the sixteenth and seventeenth centuries, may well have been set in motion by human epidemics and the subsequent demographic collapse of indigenous Americas that led to regrowth of New World forests and accelerated absorption of atmospheric carbon dioxide. These and other observations support the claim that humans have been altering biological and physical processes at the planetary scale since long before the dramatic industrial developments of the twentieth century.[23]

But palaeoarchaeologist Nicole Boivin goes further, believing that *Homo sapiens* has been altering the distribution and diversity of Earth's organisms for much more than a few thousand years to extend back at least to the late Pleistocene of more than 100,000 years ago. As human populations expanded out of Africa, they brought with them their commensals and parasites, seeds and stowaways, domesticated animals and favourite cultivars. Some of these co-migrating species came to occupy habitats and climates much like those of their origins, but others were able to construct agreeable niches out of the environments that they moved into, such as the grey rat, which radiated out of East Asia to inhabit towns and cities wherever humans built them, or the Asian tiger mosquito (*Aedes albopictus*), which came to occupy tropical and semi-tropical regions across the globe. Since human-mediated species dispersals have been happening ever since our species moved, there is good rationale for dating the start of the Anthropocene to a time long before such Earth-altering events as the invention of the steam engine, the harnessing of oxen to plough, the controlled use of fire or even the human dispersal into Eurasia and beyond, for people have been altering Earth ever since we have been human. If one envisions the world before humans existed, one will realize that all species alter the world in their own ways, with it being difficult to single out any one organism as the primary shaper of any age. It could be our own inability to escape our own ego that prevents us from finally eliminating the notion of an Anthropocene altogether – to replace it, say, with the age of interacting species, since each species produces tangible earthly changes. Any Anthropocene concept may obscure the fact that the advent of any new life form, from nucleic acid to the noosphere, produces a new age on Earth.[24]

Lynn Margulis recounts the story of how she came to agree with a student in her class who once quipped that 'Gaia is just symbiosis as seen from space.' After staring into microscopes much of her life, Margulis zoomed out to view the whole Earth as an interdependent system, with living and non-living parts relying on each other. Even if Gaia is not itself a bundle of organelles floating in cosmic cytoplasm that reproduces and undergoes planetary selection, the parallels between Margulis's microcosm and her macrocosm are so striking that she became convinced that microbiology is linked to cosmology. Symbiotic relationships play crucial roles in both sciences, with parasitic antagonists sometimes morphing into cooperative mutualists. Symbiogenesis, a term Margulis borrowed from Konstantin Mereschkowski's more speculative counterpart, denoted the transformation of parasite into symbiont to join with the host and form a single organism, thereby providing quantum leaps in evolution. She noted that such symbiogenetic events could even be witnessed in the laboratory, as in one experiment whereby a parasitic organism was found to transform itself over time into a vital bodily part. Parasitism multiplies the diversity of life.[25]

Gaia's macrobiome therefore includes hierarchies of organisms, from complicated multicellular organisms such as people, to the eukaryotic and procaryotic cells that compose them, to all the unicellular organisms that live separately but dependently in the lithosphere, hydrosphere and atmosphere. The conviction that blankets of microbes helped fashion a Precambrian planet into our current biosphere was the starting point for an early series of conferences that Margulis participated in about the origins of life. To a group of mostly physical scientists convened at Princeton in 1968, Margulis explained that 'it seems very clear that a blue green alga, that itself evolved much earlier, got inside of a non-photosynthetic cell. The chloroplast, therefore, can be considered a cell inside a cell.' This photosynthetic organelle was in turn responsible for producing much of the planet's atmospheric oxygen. Margulis went on to present the results of an investigation in which oxygen-dependent cells could be made to survive under oxygen-deprived conditions like those of the early Earth, when organisms did not rely on mitochondria to metabolize oxygen. The implication here was that the mitochondria that we moderns all depend on had long ago morphed from external antagonist to internal mutualist for allowing respiring life to flourish under oxygenated conditions. Here again, parasite had transformed into symbiont for promoting life as we know it.[26]

It was at these Origins of Life meetings where Margulis first came to know Lovelock, whose presentations likewise aimed to show how living processes could create the Earth's unusual atmosphere. Nitrogen in our air likely had biological origins, too, Lovelock explained: 'There is good reason to believe that if there were no life on Earth, there would not be any nitrogen in the atmosphere.' Their arguments joined cell with planet. Reflecting on the magnitude of these insights was astronomer Carl Sagan, also in attendance, who remarked that 'I don't believe there is any subject which is as exquisite in the interlocking of disciplines as this one.' Yet the elephant in their conference room was not whether life, once it bubbled out a cosmic soup, could develop into a string of multi-cellular life forms, for Darwin had much earlier provided powerful arguments for explaining that process, but whether human beings could upset and corrupt the essential conditions that supported life on Earth. 'The planet will not permit our populations to continue to expand,' Margulis would later warn. 'The tropical forest trees are humming to themselves, waiting for us to finish our arrogant logging so they can get back to their business of growth as usual. And they will continue their cacophonies and harmonies long after we are gone.' Sagan would offer his own vision of humanity's role in the planet's functioning, declaring at one point in his popular television series that 'From an extraterrestrial perspective, our global civilization is clearly on the edge of failure and the most important task it faces is preserving the lives and well-being of its citizens and the future habitability of the planet.'[27]

Cancer, Parasite – or Pathogen?

Marine biologist Daniel Pauly thinks big when it comes to managing ocean resources. Staring out to sea from his office in the Canadian Pacific, Pauly is convinced that most of the world's fishery catches are much higher than officially reported. Much of his life's work has focused on documenting humanity's dramatic exploitation of the world's oceans and the dire consequences that are resulting. His and his colleagues' data indicate a shocking global decline in ocean biomass over the past decades, representing an aggregate loss of more than half of what swam in the oceans in 1950. In reflecting on these declines, Pauly asks pointedly whether *Homo sapiens* should be considered a cancer or a parasite on the Earth. The transition of humans from being part of nature to that of global nemesis 'can be seen as a frightening analog of the progression of cancer tumors in someone's body.' As a good biologist, he points out that cancer cells have little ability to alter their course and will eventually kill themselves by killing their host. But parasites, for their part, 'are evolutionarily smarter . . . They may be very infectious at first, but usually a strain will select out which can co-exist with the host and may in fact turn into a symbiont.' Emphasizing the similarities between cancerous growth and humanity's rabid consumption of natural resources, Pauly wonders 'whether it will be possible to turn us humans into benign parasites on the surface of the earth . . . or whether we will continue to be part of the Earth's ecosystem in the same way that a malignant tumor is – never for a long time – part of a person's body'. [28]

As it turns out, comparing humans to cancers or parasites in an earthly system has a long tradition. In 1955 the Rockefeller Foundation's medical sciences director, Alan Gregg, warned that human population growth was not unlike cancerous growth, asking whether 'colonization of the Western Hemisphere [can] be thought of as metastasis of the white race?' Taking up these ideas was ecologist and author of *The Population Bomb* Paul Ehrlich, who warned in 1968 that 'We can no longer afford merely to treat the symptoms of the cancer of population growth; the cancer itself must be cut out. Population control is the only answer.' Adopting a similar perspective when describing urban growth, environmental writer Scott Russell offered that 'If the phrase sustainable growth means perpetual expansion, then it is a delusion … Cancer shows that rampant growth soon becomes malignant.' For his part, physician and abortion rights advocate Warren Hern felt it useful to extend this cancer metaphor, arguing that

> the human species as a whole now displays all four major characteristics of a malignant process: rapid, uncontrolled growth; invasion and destruction of adjacent normal tissues (ecosystems); metastasis (distant colonization); and dedifferentiation (loss of distinctiveness in individual components). We have become a malignant ecopathologic process. If this diagnosis is true, what is the prognosis? The difference between us and most forms of cancer is that we can think, and we can decide not to be a cancer. [29]

In the world of biological metaphors, it seems that cancers are worse than parasites. Cancers, after all, are once-healthy cells and tissues that switch their allegiance to begin threatening the larger system of which they are a part. Yet parasites, even if they also inflict damage, may carry out vital roles by becoming beneficial symbionts, as Daniel

'General-purpose parasite human'
proposed for the horror video game
Resident Evil 4, 2023.

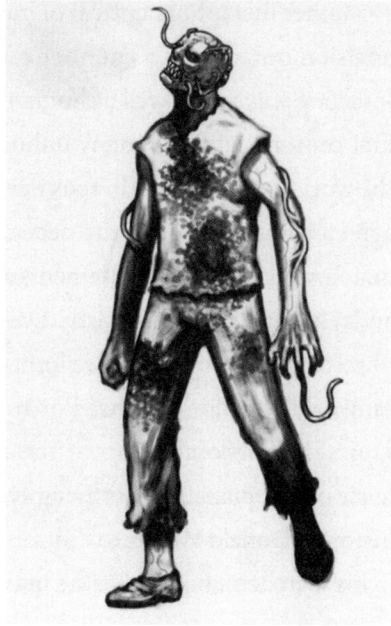

Pauly points out. There are certainly plenty of casual observers who insist that parasites are simply evil organisms without acknowledging their more subtle roles in being able to contribute to the well-being of their host. With a bit of literary flair, sociologist Sing Chew begins his *World Ecological Degradation* (2001) by asserting that 'the history of civilizations, kingdoms, empires, and states is also the history of ecological degradation and crisis. Such a historical trajectory of human "macro parasitic" activity has occurred at the systemwide structural level for at least the last five thousand years.' Evidently borrowing from William McNeill's terminology without acknowledging a parasite's more complex roles, Chew laments that many states and civilizations pursue 'productive and consumptive levels that are extremely macro parasitic in nature'. Yet one might remember that in ancient Greece, where parasites originated, they were welcomed in by their hosts precisely because they could offer desirable or useful services and would continue providing them, else their give-and-take relationship would soon end. There are, as seen, innumerable examples of how successful *non*-human parasites can provide various benefits to their hosts.

Other metaphors critical of humanity's place in the natural world envision our species as neither cancer nor parasite but as pathogen. Inflicting *pathos* may well be humanity's principal activity in our twenty-first century, with the many billions of us possibly doing just that to the world's ecosystems. In 1995 environmentalist David Brower, while agreeing that cities are quite necessary and can even be beautiful, ultimately viewed human settlements as a disturbance on the natural order and a 'plague upon the Earth'. David Attenborough echoed Brower's phrase, adding that 'either we limit our population growth or the natural world will do it for us.' For Brower as for Attenborough, *Homo sapiens* had become a sort of *Yersinia pestis*, inflicting pain and pestilence on the planet and its living systems. Building on this argument is historian Donald Worster, who combines pathogenesis with parasitism in his own depiction of Earth's main affliction:

Let us acknowledge too that we humans are also a kind of virus. We too have become epidemic on Earth. Viruses 'want' only one thing: to find a host, attach and grow, reproduce, spread, and increase in numbers. How is that different from what humans do or want? We want to reproduce and increase our numbers, and the way we do that is to become parasites on others. We have emerged through genetic mutation and, wonderfully, we have survived, spreading to all the continents and finding plenty of 'hosts' that we force to support our drive to reproduce. It is precisely because of our biological nature that we find ourselves the architects of our own tragedy.[30]

James Lovelock likewise reflects these anthro-pathogenic sentiments in his own interpretations of the main threats confronted by Gaia, but he ultimately offers more hope in humanity's potential restorative role. In 'The People Plague', a book chapter appearing on his 72nd birthday in his *Healing Gaia*, Lovelock envisions human beings as more than just another member of Gaia, having developed into a serious impediment to homeostatic harmony. Lovelock then goes on to out-line four possible outcomes of a pathogen infecting a host: in the first outcome, an invading organism may simply go on to destroy its host. In the second, the invading organism may, after some struggle, finally be destroyed by the host. The third possible host–pathogen encounter, explains Lovelock, is 'a long war of attrition' resulting in a chronic and unresolved infection. This desperate struggle seems to be humanity's current state of affairs, with people inflicting serious illness on its Gaian host at the same time the host occasionally strikes back at the pathogen – much like the scenario of a concerned doctor inspecting an infirm Earth to finally pronounce: 'I'm Afraid You Have Humans'.

But Lovelock's main reason for distinguishing different outcomes of an infective pathogen was to highlight a possible fourth relationship that might result; namely, mutualism: an encounter between antag-onists can lead to a kind of reciprocal altruism whereby each helps the other. Lovelock had humans in mind when he wrote that 'the invader is restricted to a region or role and there supported, protected, and nourished, in return for some service to the host.' It was thus symbi-otic mutualism that presented *Homo sapiens* with their best and most viable option for inhabiting Earth, according to Lovelock. 'As intelli-gent microbes,' he implored, 'we have the advantage of knowing the risks of failure and the lasting benefits of symbiosis. But will we achieve it?' For Lovelock, as for a cadre of like-minded macrobiologists, it is more hopeful to envision our species not as a pathogen or as a cancer, but as parasite transitioning to mutualism by adjusting amicably to its host. This suggestion that parasite and host can and will cooperate with each other to form a relationship benefiting both had been laid out since the days of Peter Kropotkin and the first hypotheses about declining virulence. Whether or not it is 'individual humans with their microbial diseases, or planets infested with people', concluded Lovelock, 'it is the sentient partners, the people, who need the will to live with their partners in symbiotic harmony. Whether humans are the host or the parasite seems to matter less.'[31]

Becoming a Gaian Mutualist

If symbiotic cooperation is indeed crucial to humanity's long-term survival on the planet, we need to understand the cooperative nature that may exist between ourselves and our fellow creatures. Successful interspecies cooperation centres on whether humans are taking more than they are giving to the whole Earth system. If our species has become a sort of Gaian parasite, as Lovelock and Margulis imply, is this because our species is a relatively recent arrival on Earth that began by inflicting harm on its host? One wonders whether human beings will ever be able to manifest true cooperation with their Gaian host, or if they are destined to continue aggravating their living planet. If Hans Zinsser was correct in asserting that 'Life on Earth is an endless chain of parasitism,' perhaps Gaia's principal parasite will finally learn to adjust its methods for offering reciprocal benefits to the larger support systems upon which it depends.[32]

Pulitzer Prize-winning René Dubos was convinced that human beings have indeed been developing more harmonious relationships with Earth, despite occasional setbacks that place people at odds with the natural world. Dissenting from the warnings of his contemporary Rachel Carson, for example, who at one point asked 'whether any civilization can wage relentless war on life without destroying itself', Dubos taught that 'Human interventions into nature can be creative and indeed can improve on nature.' Despite the constraints posed by a finite planet faced with shrinking resources and rising human numbers, the

> reciprocal interplay between humankind and the Earth can result in a true symbiosis – the word symbiosis being used here in its strong biological sense to mean a relationship of mutualism so intimate that the two components of the system undergo modifications beneficial to both.[33]

The suggestion that humans can improve and enhance an earthly system is certainly a very old notion. Three hundred years ago, French naturalist George-Louis Leclerc, comte de Buffon, saw humanity's labour as producing advantageous changes to the natural world. 'Primitive nature is hideous and dying; I, I alone, can make it living and agreeable,' he once declared:

> Let us dry these swamps; converting into streams and canals, animate these dead waters by setting them in motion . . . Herds of bounding animals will tread this once impracticable soil and find abundant, constantly renewed pasture. They will multiply, to multiply again. Let us employ the new aid to complete our work; and let the ox, submissive to the yoke, exercise his strength in furrowing the land. Then it will grow young again with cultivation, and a new nature shall spring up under our hands. How beautiful is cultivated Nature when by the cares of man she is brilliantly and pompously adorned![34]

Buffon was clearly echoing Christian doctrine that implored people surrounded by the Garden of Eden, to 'work it and take care of it'. Other spiritual traditions cherished similar values, with Islamic teachings, for their part, proclaiming that 'The world is beautiful and verdant, and verily God, the exalted, has made you His stewards in it.' At one point Buffon even suggested that humankind could 'alter the influence of its own climate, thus setting the temperature that suits it best'. The possibility that the planting of crops can multiply local rainfall would not have surprised the naturalist, since he felt that cultivation begat improvements at many levels. Dense forests were impediments to human progress, so that by clearing them, one brought sunlight to the soil while allowing people to expand their crops and spread civilization. Domesticating animals multiplied human labour for converting waste into bounty, with the yeoman farmer producing smiling fields of productivity.[35]

If we ask how far humans have been able to implement Buffon's vision of gardening the planet, one tabulation shows that in the time since *Homo sapiens* have multiplied and migrated across Earth, more than a thousand plant species and at least fifty animal species have become domesticated in our landscapes to flourish under human management. To this list of tamed species could be added our bodily microsymbionts, from fungi and bacterial strains to the eukaryotes and arthropods that have come to live on or in us while depending on us and we on them. Wheat and rye, bananas and yams, horses and hogs, goldfish and silkworms, farmed salmon and honeybees, together with anthropophilic mosquitoes, numerous lactobacilli and gut bacteria, as well as *Plasmodium vivax*, would all not exist in their current forms or current ranges – or at all – without the ongoing environmental modifications stemming from human activities.

To offer a quantitative overview of the earthly changes due to our species, humanity contributes the second most biomass of any mammal on the planet, and raises the mammal that contributes the most biomass, the cow, alongside the most numerous bird, the chicken. Although the human stewardship of these and many other domesticates have displaced wild creatures while transforming their habitats and accelerating species extinctions, the Earth's total mammalian mass has reportedly increased under the hand of *Homo sapiens*. Estimates place the number of cows now inhabiting India at 307 million – one for every four to five humans – and the number now inhabiting North America at 103 million, roughly double the number of bison that once roamed that continent. Even if in the last millennia the net global biomass of wild mammals is estimated to have decreased sixfold, Earth's much heavier biomass of domesticated mammals (including human beings) has increased fourfold. In viewing the Green Revolution, optimists such as Norman Borlaug have claimed that global cereal yields in the second half of the twentieth century have risen several fold. Maybe our species really has acted as a Gaian mutualist by rearranging organic matter to accelerate the earthly metabolism and harness a greater fraction of the Sun's energy for producing more life on Earth. Graphs showing rising levels of atmospheric carbon dioxide are mirrored by those showing dropping levels of oxygen, which might be explained not just by the combustion of more hydrocarbons, but by more respiration by mammals digesting foodstuffs. If humanity is borrowing from the future by consuming extra oxygen, our species is also increasing Earth's total mass of mammals.[36]

But the cautionary rejoinder to such optimistic numbers is that humanity has also engineered a significant *loss* of plant biomass over the last centuries – primarily in the form of deforestation – with forests reportedly shrinking by one-third to one-half since our species expanded out of Africa. Because vegetation makes up more than four-fifths of the planet's total terrestrial biomass, the final balance sheet indicates that human beings have served to diminish Earth's aggregate of living organisms by a great deal. In taking to heart Buffon's advice of clearing forests for spreading civilization, human populations have been sacrificing plants (as well as fresh and saltwater creatures) for the benefit of terrestrial mammals, ourselves included. With less oxygen issuing from photosynthesizing forests, less atmospheric carbon dioxide is being reabsorbed, thereby heightening greenhouse effects. Gaia's gaseous cloak is reaching carbon dioxide concentrations never before witnessed by people. Measured by changes in the atmosphere and changes in biomass, it appears that *Homo sapiens'* recent effects on the planet have been negative and disruptive. Perhaps we really do need to take to heart George Perkins Marsh's warnings that 'Man is everywhere a disturbing agent.' Perhaps Vladimir Vernadsky spoke the truth when surmising that

> Civilized humanity has introduced changes into the structure of the film on land which have no parallel in the hydrosphere. These changes are a new phenomenon in geological history, and have chemical effects yet to be determined. One of the

233

principal changes is the systematic destruction during human history of forests, the most powerful parts of the film.[37]

And yet, before branding humanity as a scourge on the planet, one might still envision the day when atmospheric carbon dioxide levels can be regulated though technical as well as social means, such as by injecting liquified greenhouse gases underground at the same time as convincing people to consume less fossil fuel. Foodstuffs might be grown in greater quantities to be abundantly and more equitably distributed around the world. All peoples might gain access to adequate sanitation along with the miracles of modern medicine. Perhaps humanity really can usher Gaia towards a leaner but healthier geophysiological state, one that favours humans and their domesticates, but also one that grows abundant, carbon-absorbing forests. Even if human activities have contributed massively to our day's sixth mass extinction episode while commandeering many of the world's pristine ecosystems, our species has also gone a long way towards constructing a global garden that will sustain billions while providing habitat for our mutualist co-travellers. As Marsh himself qualified:

> The physical revolutions thus wrought by man have not all been destructive to human interests. Soils to which no nutritious vegetable was indigenous, countries which once brought forth but the fewest products suited for the sustenance and comfort of man ... have been made in modern times to yield and distribute all that supplies the material necessities, all that contributes to the sensuous enjoyments and conveniences of civilized life.[38]

It is possible that Gaia's principal antagonist may still employ amazing ingenuities to amend for its lapses of selfishness and greed. If, in prehistoric times, humankind once lived as a mutualist in relative harmony with nature, to then fall from Gaia's grace and begin disrupting and parasitizing the earthly garden, there are those like Dubos, Buffon and Marsh who maintain hope that we can again become mutualists – or, as Gregg Easterbrook subtitles his new book, *Reasons for Optimism in an Age of Fear*. Pierre-Joseph van Beneden taught that even if a parasite's profession is to live at the expense of its neighbour, it does so cautiously and cleverly so as not to endanger its patron's life: 'The parasite does not kill; on the contrary, he profits by all the advantages enjoyed by the host on whom he thrusts his presence.'[39]

Biospherics

In searching for better ways for humans to contribute to a healthier Gaia, one should finally mention the man from Cranleigh, a town in Surrey near London, who claims to have assembled the world's longest living terrarium. In 1960 David Latimer dropped a bit of soil into a ten-gallon, narrow-necked bottle before adding some seeds of spiderworts and then sealing the contraption, with the resulting self-enclosed jungle growing and thriving ever since. Only once did he pull the cork in 1972 to sprinkle in a little more water, but otherwise nothing else has entered or exited this mini-ecosystem, except for a daily input of sunlight. It turns out that Latimer may be conducting the longest known experiment with an artificial biosphere, a system that has been manufactured in many versions in the hopes of recreating approximations of our larger Earth system. Vladimir Vernadsky might agree that the best way to begin to understand our biosphere is to set out to recreate one in miniature for observing how long the system can survive or how far it might be able to regulate itself.[40]

Terraria have a rich past, with the first enclosed plant vases made famous by Nathaniel Bagshaw Ward, an English doctor and botanist who invented them in the early nineteenth century. Ward found that plants living within sealed glass containers grew remarkably well, sheltered from dismal British winters where cool temperatures and sparse sunlight meant that most exposed plants withered or hibernated,

Wardian case, 1852.

particularly ones imported from warmer climes. Wardian cases, fashioned into various shapes and sizes, found wide use on ships for
transporting distant flora, serving to shelter these exotic plants during
their journeys to temperate climes. As a tiny sealed greenhouse that
required almost no maintenance, the Wardian case would go on to
adorn many Victorian parlours while also serving to transport sundry
alien (and possibly invasive) species from the New World back to the
Old. Gardeners observed that Wardian cases might be permanently
airlocked to go on incubating their living payloads for years or decades,
depending on the composition of plants and creatures placed inside.
Garden historian Christopher Thacker judges the Wardian case to be,
along with the ha-ha and lawn mower, 'one of the great inventions in

garden history'. In this way, the terrarium was born and with it a kind of Lilliputian biosphere.[41]

At the same time Latimer was sealing his famous terrarium, Russian cosmonauts were developing their own live-in terrarium, which they called BIOS-1, for commemorating as well as testing Vernadsky's revolutionary ideas. With the competitors in the space race eyeing the moon and beyond, one immediate challenge was to make sure that human beings could survive extended spaceship travel. Oxygen might be carried aboard, as could food, but both would run low after a few days or months unless more of these provisions could be generated during the voyage. Completed in 1965, their multi-cubic-metre terrarium based in Krasnoyarsk, Siberia, found a human being subsisting inside on oxygen generated by a large carpet of *Chlorella* algae kept alive by a bank of xenon lamps. Thereafter, BIOS-2 and then BIOS-3 were more ambitious, follow-up projects designed to fine-tune the optimal threshold of photosynthesizing algae needed to keep humans breathing while adding edible plants such as wheat and vegetables that were useful for human consumption. Between 1972 and 1984 various crews of two to three members living for months at a time within the underground BIOS-3 were able to generate most of their oxygen and recycle much of their waste and water while growing much of their food. The cosmonauts may have had their sights on the stars, but they were really field testing the limits of life on Earth by revealing the roles that various organisms, including humans, would need to play in order to support an interdependent closed community. By building artificial biospheres and determining which creatures could live within them, people were aiming to become masters of their own destiny. Could humans act as gods, and even become good at it?[42]

The American answer to the live-in terrarium was constructed in Oracle, Arizona, following on the heels and advice of their Russian colleagues, even borrowing their label. After nearly a decade of construction, Biosphere 2 was christened in 1991 as the world's second complete biosphere modelled on Earth itself, or Biosphere 1. The sealed glass, steel and cement behemoth included more than 3 acres of sunlit living space along with 3,000 species of plants and animals grouped into seven biomes, complete with an ocean and coral reef system, together with eight human beings who would embark on their own biospheric journey over two full years. Their goal within their desert bubble was to grow all their food and recycle all their vital gases while utilizing all organismal wastes to approach complete autonomy. Their system was meant to depend only on the solar energy beaming into their facility – excepting, as was revealed later, some supplementary power from a backup generator running off fossil fuels. To much public fanfare, the Biospherians hoped to demonstrate that humans could live apart from the Earth, and in close quarters for extended periods, as would be the case during deep space missions, while fine-tuning the conditions and organisms necessary for producing an optimal and self-sufficient artificial closed ecosystem – which would also serve as a prototype shelter in a post-apocalyptic Earth. As Margaret Augustine, one of the directors of the project, explained, 'I think that the development of the Biospherians in Biosphere 2 is going to be a gauge or a tool to train Biospherians for Biosphere 1.' John Allen, the eclectic

Biosphere 2 viewed from the inside, 2021.

	EARTH	BIOSPHERE 2
Percentage of surface area that is ocean	71%	15%
Maximum depth of ocean	11 km	7.6 m
Maximum height of gaseous atmosphere above land surface	17 km (extent of troposphere)	23 m
Ratio of biomass carbon to atmospheric carbon	1:1	9:1
Residence time for atmospheric CO_2	8–10 years	2–4 days
UV radiation	Some received at Earth's surface	Glass filters out
Community size	Relatively large	Relatively small
Species density	Relatively low average species diversity/unit area	3,000 species housed in 1.28 ha.

Comparison of Earth (Biosphere 1) with Biosphere 2 (from John Peterson et al., 'The Making of Biosphere 2', *Restoration and Management Notes*, x/2 (1992), p. 165).

mastermind behind the project, saw special significance in Lovelock's and Margulis's theory of the biosphere as a biogeochemical, autonomous, self-regulating system, with Biosphere 2 becoming a 'critical experiment' to verify that theory. Biosphere 2 was therefore envisioned as an ultimate test of sustainability – 'a metaphor for the planet and our inhabitation of it' – with each plant, animal and human being needing to maintain the larger system's interests above those of any single member. Mutualistic symbiosis was the rule, else the system would collapse. Defectors and parasites could not be tolerated.[43]

The actual story over those two years of symbiosis – together-living – reveals that behind the smiles refracting through the glass ark's windows and flashed on television screens around the world, oxygen levels began sinking dangerously low, a crew member required temporary medical evacuation, luxury items were smuggled inside and group harmony soon dissolved. Even worse may have been the various canker blights appearing on the in-house crops because these and other pests had been unknowingly transported into the facility despite meticulous preventative scrutiny. Species of cockroaches and crazy ants, invasive and problematic, likewise began multiplying within this enclosed world. Philosopher Luc Peters interprets the arrival of pests to the facility as having been 'Two years of creating an artificial world cleansed of all the unwanted ingredients and guests. However, despite all the good, clean and scientific intentions, parasites intruded,' he explained, 'eating the crops, sucking up the oxygen, intruding in an incomprehensible and secretive way.' Echoing Michel Serres, Peters surmised that here, as in all systems, 'the parasite always intrudes.' Despite auspicious beginnings, the biotic microcosm of Arizona's life-sized terrarium was spinning out of control, suffering from the setbacks of surprises and selfishness, to begin unravelling under its own hubris.[44]

The creators of Biosphere 2 had announced that humanity must learn to be 'a creative collaborator with the biosphere, rather than a parasite weakening the host'. One of the inspirational advisors on the project, systems ecologist Eugene Odum, had counselled that 'Since man is a dependent heterotroph, he must learn to live in mutualism with nature; otherwise, like the "unwise" parasite, he may so exploit his "host" that he destroys himself.' Two of the Biospherians, Abigail Alling and Mark Nelson, reflected that their glass mansion might really be called Noosphere, as their Russian colleagues had once suggested, 'because of the necessity for humans to act intelligently inside the miniature world with an environmentally designed technosphere and thus make a harmonious cooperation with life.' Even James Lovelock held out hope that humans might play a positive role in managing the larger Earth system, writing near the end of his long life that 'I like to think of humans with their intelligence as . . . a life form that is very dangerous yet with near infinite promise; we are Gaia's gamble for a more secure old age.' During these optimistic moments it seemed that, with insight and foresight, the human species might just contribute to Gaia's greater longevity.[45]

The invitation cover to join the eight Biospherians in celebrating
their 'Re-entry to Earth's Biosphere' on 26 September 1993.

But in all these experiences with mock biospheres, small-scale biospheres and simulated biospheres, humans have so far not shown themselves to be the conscientious mutualists they had hoped to be. With
Biosphere 2's crew finally exiting their grand Wardian case in 1993,
having persevered during their last months with a bit of supplementary
fresh air and auxiliary fossil fuel, along with a few other odds and ends
slipped past the air locks, they would come to blame their artificial
atmosphere's troubles on their glass ark's organic soils consuming too
much oxygen combined with its concrete superstructure absorbing
too much carbon dioxide. Yet even if they had been able to optimize
their biosphere's gas composition, there was still the lurking problem
of self-interest overpowering the common good, along with far too
many unknowns interfering with the life-support network of a closed
ecological system.[46]

In the end, humans had constructed their biospheric niche in their
own image without fully understanding, and possibly never being
able to understand, all the variables involved in properly managing it.

The eight-member troupe who had entered Biosphere 2 complete with matching orange bodysuits, and who had been able to survive two years with a little help from the real biosphere, had once labelled themselves 'The Theatre of All Possibilities'. Yet when the experiment finally came to an end, after all the reports were submitted and all the diaries were closed, at least one critic maintained that the real lesson to be learned from this grand adventure was about 'the limits to earth's possibilities'. It might be true that *Homo sapiens* can contribute to a more agreeable and longer-lived Gaia. But our current experiences living within an actual global biosphere demonstrate that we are still honing our mutualistic responsibilities.[47]

Restoring Parasites and Rewilding Microbiomes

Nature is the best manager of natural processes.
From the website Rewilding Europe, 2025

Terra Preta, the curious dark soils found across large swathes of the Amazon, are thought to be the product of ancient peoples mixing, kneading and curing the ground beneath their feet to make it as productive as possible. Alongside the fragments of charcoal, ceramics and other human artefacts found in these soils are multitudes of tiny creatures that contribute to and depend upon these substrates, having perpetuated these Anthropocenic formations for at least the last 6,500 years. Biologists reveal that *Terra Preta*, which can be a metre deep in places, houses up to 25 per cent more bacterial varieties than nearby relatively pristine soils, along with almost four hundred species of fungi, protozoa, roundworms, earthworms and arthropods not found in the unmanaged soils. In stewarding the Earth, humans have since long ago made these areas more biodiverse and more productive.

Only recently have modern land uses and exploitative agricultural practices served to diminish this abundance, ushering in calls to restore the soil's former biological diversity and bring back some of the richness that once was. Pellets of this Amazonian dark earth can even be mixed into impoverished soils elsewhere to begin multiplying microorganisms and boosting fertility to grow hardier crops and healthier forests. These inoculated soils also become more 'climate smart' by capturing more carbon while deterring certain crop pests and recycling local organic wastes. Just as faecal transplants can transfer healthy microbes from one person's lower intestine to another's to begin alleviating various auto-immune diseases in the recipient, *Terra Preta* treatments inject crucial microbes into depleted soils to initiate the process of what one company calls 'regenerative reclamation'. Such experiences demonstrate that humans do not always act harmfully and parasitically on biophysical systems but can help restore and regenerate them, even on very large scales. 'We must abandon the politically and psychologically loaded idea that the Anthropocene is a great crime against nature,' James Lovelock once declared, emphasizing the symbiotic and mutualistic potential of our species.[1]

Efforts to rejuvenate soil microbes are just part of the louder call to restore and rewild Earth's living systems. In another instance of Gaia-scale restoration, the goal is to maintain or return tundra microbes to their frozen, dormant state so that they will generate fewer quantities of problematic greenhouse gases. Ecologists Sergey Zimov and his son Nikita have become famous for their project in northern Siberia that rests upon the idea that melting permafrost is a leading accelerator of climate change. Up to hundreds of metres deep, the frozen formation that is permafrost houses a spectrum of bacteria, algae, fungi and archaea. If temperatures rise above freezing and water droplets form in the permafrost, as they do every summer in parts of the upper, sun-warmed layers, the microbes that thaw there kick into action, released from their inanimate state to begin producing carbon dioxide along with more problematic methane, depending on how much oxygen is available for their metabolic processes. The aim of the Zimovs's 160 square-kilometre 'Pleistocene Park' is to reinstate widespread grazing and trampling for beating back arboreal vegetation and recreating grasslands that will once again allow the winter's cold temperatures to penetrate deep into the permafrost. The ecological rationale here is that once-plentiful woolly mammoths (and a few other tundra grazers now extinct) can be substituted by modern herds of musk ox, bison and horses. The hope is that restoring age-old arctic grazing will in turn help restore atmospheric health – even if recent observations show that the process may take much longer than expected. As with efforts to re-cultivate healthy soils, proper microbial management is at the core of this project to maintain and restore permafrost to its frozen state.[2]

If we turn to our own veins and intestines, scalps and skin, as well as to those of our animal and plant friends, we can discover still more imbalances and impoverishments in microbial diversity, as well as more efforts to bring back their healthy states. Our ongoing exposures to toxins and antibiotic treatments, our diets made ultra-hygienic through extensive refrigeration, cooking and pasteurization, our hours spent in aseptic homes and workplaces mean that our bodies are no longer hosting as many organisms as they used to, or as our parents' and grandparents' bodies did. Although our day's ease of rapid and frequent travel and varied foodstuffs is exposing us to new kinds of viruses, bacteria, protozoa, archaea, fungi and arthropods, such creatures are increasingly the same ones found worldwide and are being remixed on a global scale, so that humanity's total number of symbionts is shrinking, with one researcher estimating that a third of our primordial bodily creatures have now disappeared. Tabulations show that the mere act of immigrating to the United States, for example, leads to a loss of one's microbiomal diversity; likewise moving from country to city, as in Senegal, also impoverishes the body's microbiome, especially in infants. Our current sixth mass extinction episode is happening in our jungles, prairies, arctic and oceans, just as it is happening inside ourselves and our domestic co-travellers. Although any disappearance of our most pathogenic parasites would seem to signal only improvement and human progress, we may also be reaping crucial advantages even from these most evil of microbes, for they can keep our immune systems strong, protect us from more virulent microbes, sicken our enemies, assuage our allergies or just possibly (over several generations) promote more social equality, as explored earlier. Other, more mutualistic parasites are also disappearing in our bodies and across the planet, so that a looming challenge is now to bring back some of these internal creatures. Like the efforts to restore microbial variety to soils and large ungulates to the tundra, there are good reasons for bringing back diversity to our insides. What has been our experience at restoring parasites and rewilding microbiomes? If it is true that many parasites, from the microbial to the macroscopic, provide crucial benefits to our bodies, our ecosystems and the whole planet, then the project of restoring them requires thinking holistically about parasitism, and about how parasites and hosts interact with each other and with their larger communities.[3]

Endangered Parasites

Although the disappearance of some parasites is a normal and ongoing evolutionary process, whereby more fit or more fortunate species replace those currently inhabiting their hosts, our current accelerating numbers of parasite extinctions can be traced mostly to human activity, directly or indirectly. Just as large, free-living creatures around the world are threatened by the usual suspects of habitat loss, species invasions, pollution and climate change – being the main causes of extinction – so too are internal micro-organisms subjected to the these special challenges of the Human Age. Indeed, there are strong indications that the world's parasites are even more susceptible to our day's rapid environmental changes, since smaller co-dependent organisms are typically more sensitive to slight alterations in ambient temperature, chemical composition and biotic interactions. Conservative estimates of the rates of parasite extinction in all creatures over the next fifty years are pushing well into the double digits, anywhere from 5 to 29 per cent extinction rates, depending on the class of parasites involved. Of the world's creatures that are divided into 30 to 35 recognized phyla, at least ten such phyla are composed predominantly of parasites, and a good many of these are endangered. One study of marine fishes captured over the past century, for example, found a 38 per cent loss of their multi-host, obligate parasites, which may be due, in part, to a one-degree Celsius rise in seawater temperature over that period. There are clearly other explanations for dropping numbers of fish parasites beyond climate change, such as accumulating water contaminants, greater fish harvests and rising numbers of invasive marine species, to name a few, yet common to nearly all of these ecosystemic changes are their origins in the activities of *Homo sapiens*. Anyone concerned about the global extinction crisis needs to be especially concerned about disappearing numbers of parasites – living within and beyond humans.[4]

The higher susceptibilities of parasites to environmental stressors mean that they can sometimes be used as indicators of a host's health, like canaries in a coal mine. The presence in a host of vigorous and diverse kinds of parasites along with steady parasite numbers often signals that the host is quite healthy, whereas a sudden multiplication of parasites in a body or else a drop in their numbers can be a proxy for the declining health of the hosting partner, be it a fish or a human. There is, after all, the simple observation that human body lice may at some point begin marching systematically away from a very sick or very elderly person, with the little arthropods realizing they need to set out for greener pastures when their host begins struggling to supply them with adequate resources. In the days when lice were much more common – as they still are today in crowded and impoverished places around the world – people set out to kill lice and disrupt their habitat through simple grooming, sometimes with the help of fine-toothed combs and finger-squishing, always made challenging by the fact that each lice species tends to match the colour of its host's skin. Modern laundered clothing, along with regular treatments with shampoo and hair removal, have served to further sanitize the skin's sebaceous habitat, to the point that certain species of pubic lice, at least, are thought to be endangered. Concerted efforts at removing general body lice will certainly spare us a good deal of scratching, while also providing some defence against the real concern of contracting a number of serious lice-transmitted diseases – yet wholesale de-lousing of much less dangerous pubic and head lice may also eliminate these creatures' other services or advantages still unknown or undiscovered by us. Repeated rubbing on of pyrethrins, spinosads, permethrin, lindane, malathion or ivermectin may serve to exterminate all human lice species, but doing so also exposes us and our world to a medley of dangerous poisons, exacerbates the stigma associated with carrying the little creatures and eliminates opportunities for immune system development. One wonders if it is time to make at least some peace with our intimate arthropod by allowing a few of them to return to their homeland or even make it more habitable.[5]

Another place that might be made more inviting for our intimate co-travellers is the human mouth. Although there are no extensive historical databases for tracking the numbers and varieties of creatures that have lived between our upper and lower jaws, there is the likelihood that we have also been compromising biodiversity there. As the second most species-rich repository in the body, surpassed only by the lower intestine, the oral cavity offers a variety of temperature-controlled, mucus-permeated habitats that lie on the gums, under the tongue, down the throat and across the tooth enamel. This 'oralome' reportedly contains around 1,000 microbial species, which varies from

Selection of head lice combs found on board the Tudor carrack Mary Rose, which sank in 1545. The fine-toothed side is for removing lice and nits; the coarse-toothed side is for combing hair.

Brushing teeth: Tsukioka Yoshitoshi, *Waking Up: A Girl of the Kōka Era (1844–8)*, 1888, woodblock print.

one person to the next so as to give everyone their own distinct breath odour. Medicated lozenges, mouth washes and toothpastes date back centuries the world over, so that it is fair to say that our current oralome is the co-evolutionary product of generations of creatures aiming to thrive in our mouths while habituating themselves to the foods we eat and dealing with the human compulsion for gargling, rinsing and tooth-scrubbing. Although almost no dentist will suggest that a lifetime of tooth brushing has damaged our oral biota, there is little question that the main differences between the oralome of *Homo sapiens* today and that of our distant ancestors depends largely on which substances have been habitually placed in the mouth, chewed on, swished and swallowed. As disinterested a company as Colgate will only say that 'Compulsive or over-vigorous brushing can lead to oral health problems and put your mouth at risk for dental abrasion, tooth sensitivity, and gum recession.'[6]

Particularly disruptive to a mouthful of healthy microbiota is chewing tobacco, as well as smoking anything, and even inadequate brushing, at least in the modern hyper-hygienic West, since certain oral strains may multiply beyond their current proportions unless reined in through periodic – but not excessive – chemical and mechanical controls. Meanwhile, one's deeper physiological temperament may also damage the oralome, with conditions such as diabetes and hormonal changes linked to imbalances in the mouth's flora and fauna. Once an oral cavity's microbiome is pushed far beyond its normal composition, serious consequences may follow, such as the accumulation of lactic acids that can lead to accelerated tooth decay or worse. Similar to the warnings arising over missing microbes across the rest of the body, an impoverished oralome is suspected of causing higher incidences of heart disease, Alzheimer's disease, even pregnancy complications and facial cancers. Although it is largely unknown how any one oral microbe's absence or presence may be causing wider systemic problems for the host – other than to acknowledge that unhealthy chemical imbalances are at the root of them all – one might remember that human beings are by sheer cell numbers less human than viral, bacterial, fungal, archaeal and protozoan. A person needs to take good care of one's co-travellers in order to feel healthy.[7]

Co-Conservation

Parasite conservation can take many forms. To bolster the numbers and varieties of creatures living inside other creatures, one immediate strategy is to protect and conserve a parasite's immediate environment, making it off-limits to toxins or creatures that are antagonistic to the daily affairs of this organism. As with nature conservation at larger scales, the first order of business is to erect barriers to encroachment from outside, in the same way that a preserve or national park establishes borders for protecting creatures within it so that these indigenous organisms can flourish on their own. Successful nature conservation may also require active human manipulation, such as by eliminating threatening species while introducing beneficial ones or by correcting attributes that are weakening the target species – for example, by providing habitat corridors between natural areas (in the case of parks) or by changing one's hygiene habits (in the case of the microbiome). But the first rule for conserving parasites is to save as much of their habitat as possible, which means saving the host's habitat, thereby requiring co-conservation.

In the Alps of northwestern Italy, red deer (*Cervus elaphus*) disappeared some two centuries ago following hunting pressure and habitat loss. Through a series of reintroduction projects over the past few decades, coupled with spontaneous migration from adjoining regions, this deer is becoming re-established in these valleys. Although a handful of parasite species were at first hard to detect in the newly returned deer, recent observations show that sundry botfly larvae and nematodes are becoming more common in the animals, thereby resulting in unintentional parasitic recovery. Since many parasites are obligate to their hosts, being unable to live independently from them, the crucial measure needed for conserving parasites is conserving their hosts. Red deer botfly larvae and red deer nematodes could not do well without red deer. And while these grisly parasites that sometimes produce oozing skin wounds on an otherwise vigorous deer would seem to merit little of our sympathy, or that of the deer themselves, a group of conservation biologists reminds us that 'not all parasites are detrimental during host conservation activities and exposure to parasites may even convey benefits to translocated individuals and populations.' Specifically, captively bred animals that are reintroduced are usually less susceptible to the harms that parasites can inflict if the animals have already had earlier exposure to them.[8]

It should be added that on the other side of the Atlantic in the woodlands of the Great Lakes region, biologists have found that forest mice infected with their own species of botfly actually go on to live longer lives than mice that are free of these irksome creatures. As with red deer, the mice find botflies burrowing into their fur, but mysteriously, it has been repeatedly confirmed that 'mice infested with bot fly larvae persisted significantly longer than uninfested mice, and those with two or more bot fly larvae persisted longer than those with only one larva.' Scratching their heads – while repeating the dictum that 'Parasites, by definition, have a negative impact on their host' – some biologists suggest that such parasites are diverting energy from a host's reproductive success into its longer life, thereby resulting in net harm to the mouse population. But the simple fact remains that, somehow, parasitic botflies serve to lengthen an individual mouse's lifespan. An alternative and more controversial interpretation here would be that over the short term, at least, the unsightly botfly and its larva are not as parasitic as supposed – or may not even be 'parasites' at all in the traditional sense, but are instead partners with their hosts in promoting longer lives in each other. Just possibly, it will be found that a botfly's short-term advantage to a mouse's longevity balances its long-term disadvantage to the mouse's reproduction, since botflies and mice have been able to co-exist with one another for so long. The point here is that for conservationists focusing on protecting biodiversity, parasites can help save hosts just as hosts can help save parasites.[9]

In some cases, there may be very few hosts left to provide a parasite with much habitat, or parasite numbers may have dropped to such low levels that they have difficulty reproducing, requiring emergency intervention through intensive restoration. Like a salvage archaeologist confronting an excavation site soon to be inundated with the rising waters of a hydro dam, a parasite conservationist may need to resort to extreme measures to save a species. The only parasite currently listed on the International Union for Conservation of Nature's Red List is the pygmy hog-sucking louse (*Haematopinus oliveri*), which has attracted such concern largely because its host, the critically endangered pygmy hog (*Porcula salvania*), made the list. Isolated to a small range on the Bhutan–India border, these charismatic mega- and mini-fauna are both now coming back from the brink thanks to an intensive programme of captive hog breeding. The best way to restore the parasite has been to simultaneously restore the host. And since this hog louse carries its own host-specific parasites and symbionts, a salvaged pygmy hog represents a multiplier factor for biodiversity conservation. Rescuing a Russian doll rescues the many dolls inside it.[10]

The project of saving parasites may resort to still more extreme measures, pushing conservation into the realm of laboratory vials and frozen repositories. But reproducing conditions inside the test tube like those existing inside the host can be challenging or even impossible. In the case of salvaging rare strains of malaria parasites, whether for study or for conservation, there is usually human blood available that will preserve a species for a time. Yet without also cycling a *Plasmodium* through its definitive host, the mosquito, this parasite's days are numbered. For this reason, live mosquitoes are sometimes cultivated and maintained alongside human blood media for reproducing a parasite's whole world *in vitro*. In the case of biorepositories, mere fragments of a parasite can be placed in a deep freezer for preservation and eventual retrieval when the need arises. The University of Nottingham is conserving 'tissues, gametes, viable cells and DNA' of endangered creatures in its Frozen Ark Project that might be thawed some day to theoretically recapture the essences of bygone parasites. In the same way, the Smithsonian has built an inventory of more than 4 million vials that maintains strands and specimens of living tissues in freezers ranging between +4°C and −190°C. Meanwhile, Svalbard's Global Seed Vault, which currently protects more than a million seed samples in the frozen recesses of an arctic cave, is also conserving traces of plant parasites in the form of fungal spores attached to seed coats, no matter how meticulously the seeds have been prepared and treated. Some of these fungi will probably spring back to life if the seeds are ever germinated. But all will not be lost for this newly resprouted vegetation, since many fungi provide critical functions for the plant world, such as enhancing a plant's ability to take up critical nutrients, even if these parasitic organisms consume some plant tissue in the process. A sample of parasites and representative parts of them may be salvaged in the southern hemisphere as well, since an Antarctic interest group is planning a biorepository of its own to be stocked with creatures found across this frozen continent. Any effort to save and restore biodiversity should not, and in some instances cannot, ignore the smaller creatures that come along for the ride – if only for the simple reason that such creatures may be compensating their hosts for that ride.[11]

Probiotics to the Rescue

The direct consumption of viable microbes is one of the most immediate ways for replenishing the creatures in our body. By swallowing probiotic pills or liquids, or slathering on creams fortified with live ingredients, the hope is to bring back our lost micro-creatures or add wholly new ones that will improve digestion, circulation, respiration and any of our other organ systems. There are also probiotics available for our dogs and goldfish, feedlots and garden plants. The concoctions meant for human consumption are typically laced with bacteria and yeast, and include such milk-product constituents as *Lactobacilli* and *Bifidobacteria*. Yet all the other normal lifeforms of our intestines and skin, from protozoa to arthropods, may be left out of these probiotic doses, which if swallowed may only travel as far as the stomach before being denatured by our harsh bodily secretions. Optimistic dietitians nonetheless advocate recreating part of what they think our bodies once were, even reinforcing them with novel creatures thought to multiply our vitality. There are certainly many reports of amazing cures brought about by consuming aliquots of living organisms – along with a good number of warnings about probiotic therapies gone bad. After all, such treatments may only resuscitate a fraction or even a harmful portion of our primordial microfauna. The u.s. National Institutes of Health twice cautions on its webpage that 'Cases of severe or fatal infection have been reported in premature infants who were given probiotics.' Meanwhile, the uk's Advertising Standards Authority has been busy reining in the health claims of certain probiotics manufacturers, as in the case of Yakult, a fermented milk company that declares that the 'friendly' Japanese bacteria saturating its products are 'strong enough to reach the gut alive'. The claim was eventually upheld.[12]

Even if 'probiotics' is a term that did not find common usage until the 1990s, clearly as a rhetorical counterweight to 'antibiotics' that have been in wide use since the 1940s, the history of the therapeutic consumption of microbial creatures might be traced as far back as when people started eating food. Not only do yoghurts, yeasty bread, fermented beverages and active cheese cultures – such as Sardinia's *casu marzu* – contain live creatures that people daily consume, but so do fruits, grains and animal parts of all kinds and of varying degrees of freshness. Raw honey, caviar, ceviche and sushi, as well as ants, grubs, termites and beetles, are side dishes or main courses around the world, and all of them contain varying degrees of micro- and macro-organisms, some of them still squirming on the plate. Thailand is a heavy consumer of crickets – baked, fried, powdered and raw – with such insects made more appetizing after one learns that swallowing them can help reestablish crucial gut fauna. Although a casual or frequent cricket eater may also risk allergic reactions, choking hazards and occasional pathogen infections, a number of studies suggest that consumers of these insects can also enjoy generally improved gut health. Live foodstuffs have long been recognized as providing more than just energy, while seeming to offer such health advantages as better digestion, heightened disease resistance and longer life. Researching in the early 1900s, the Russian Élie Metchnikoff, working at Paris's Pasteur Institute, is credited as the first to suggest that active bacteria in Bulgarian yoghurts are the main reason why the local villagers consuming them enjoyed such robust health. Metchnikoff also felt that strategic dieting would serve to replace harmful intestinal microbes with useful ones, in what can be considered an early call to restore the human microbiome. But even earlier, says Reinhard Hoeppli, an eighteenth-century French physician by the name of A. C. Lorry was already calling attention to how our tiniest bodily creatures might be improving our health, as in the case of a ripe rash of scabies (being a subcutaneous infection of mites), which could be useful 'for treatment of asthma, inflammations and malignant fevers' in a vision that predates our own call for resolving auto-immune diseases with worm therapies by three centuries. Scabies was at the time also mentioned as a useful treatment against insanity, implying that a person might cultivate mites on one's sleeve in order to reap therapeutic benefits. Although these careful observers of yesteryear may at times have been drinking too much dandelion wine, there is no denying that their notebooks are also filled with meticulously collected data for lending more credence to their claims.[13]

And even though 'animalcules' such as bacteria and protozoa could not be viewed directly until the development of the first strong microscopes in the 1600s, there was already an understanding that certain vital humours, from saliva to blood to excreta, could be consumed, removed or combined with other foodstuffs for beneficial effect. In the world of animal husbandry, farmers had early on discovered that their goats or cows that issued smelly belches or runny stools might have their digestive disorders improved by being made to swallow a bit of chew from a healthy animal nearby. In this way, one cow's bolus could become another cow's hors d'oeuvre, with a clump of partially digested grass often going on to resolve the recipient cow's indigestion. Whether or not the practice of pre-chewing a human baby's stringy meal began after a parent observed the successes in the barnyard is an open question. A group of researchers will only say that abandoning the common practice of pre-mastication, as in modern China where 63 per cent of university students report having eaten pre-chewed food as infants, 'has placed children at increased risk of inadequate nutrition and decreased ability to confront infections'. Although there is some concern that a pre-chewing mother (or father or other caregiver) could pass on such diseases as AIDS, this practice may nonetheless 'promote immune tolerance and perhaps modulate subsequent allergic responses'. Likewise the ritual of *Tahnik* carried out in some Islamic regions, whereby the chewed fruit of date palm is rubbed onto the palate of a newborn baby, may be promoting similar microbial benefits.[14]

'Transfaunation' has for many decades been the preferred veterinary term for the project of moving micro-organisms between bodies to involve a range of transferrable fluids. Italians have been carrying out mouth-to-mouth transplants in their domestic grazers since at least the late middle ages, as when anatomist Fabricius Acquapendente described and praised the technique. Not to be outdone, modern Chinese researchers point to colon-to-mouth transfers of fluid carried out in the Dong Jin dynasty of the fourth century AD, albeit in human subjects, with positive health results observed after administering 'human fecal suspensions by mouth for patients who had food poisoning or severe diarrhea'. The researchers explain that this 'yellow soup', as the mixture was euphemistically called, can be considered a direct forerunner of modern faecal microbiota transplantation (FMT), which has been able to resolve *Clostridium difficile* infections by up to 91 per cent upon a first treatment and is even more effective upon a second treatment. The flowering of *C. difficile* in a person's gastrointestinal tract is typically the result of having taken too many antibiotics, with human probiotics via transfaunation coming to the rescue.[15]

Mention should be made here of so-called *Dreckapotheke*, or 'dirt pharmacology', a term popularized in the late 1600s by Christian Paullini, a cosmopolitan European medical healer who promoted the use of various bodily secretions from urine to semen to excrement for mixing into lotions or foodstuffs to help remedy everything from skin rashes to stomach aches. Such secretions were collected from humans as well as barnyard animals, and sometimes combined with dirt and soils along with various 'filthy' waters to find scattered use among European peasantry. Petra Maurer, a historian of traditional medicine, recounts how Paullini's brand of medical healing found precedents across central Asia and Tibet, so that dirt and bodily productions combined with other materials into various microbial concoctions had long been relied upon by human and animal healers alike for bringing relief to their patients. Although later medical experts would dismiss these early 'dirt pharmacologists' as peddling barbaric folk practices, there is little doubt that their methods served to multiply microbiomal diversity, which must have exacerbated infections in some instances but may have alleviated painful symptoms and promoted healthier conditions in others. Even in the case of urine, microbiologists are now discovering so many kinds of microbes living in the clear, yellow liquid issuing from healthy kidneys that they have coined it the 'urinome'. In traditional Tibetan medicine, urine served a variety of uses, with that of eight-year-old boys typically finding special favour for treating difficult cases. The renowned Sangyé Gyatso, who founded Tibet's Chagpori School of Medicine in 1694, taught that consuming a young boy's especially pure urine 'heals plagues, contagious fever, demonic possession, poisoned bone tissue and breathlessness'. And geophagy, for its part – the practice of eating dirt – has survived up to the present under various guises across the world, often as an act of desperation when food is scarce but also serving as a microbiological and nutritional supplement. The Rockefeller Sanitation Commission campaigned against dirt eating in the southern United States because it realized that this habit favoured hookworm infection, and only much later would hookworms be acknowledged as the largely harmless helminths that they are – except among worm therapists, who now consider this roundworm as an inoculant of choice. Even if hookworm infestation is currently classified by the WHO as a 'neglected tropical disease', other health experts point out that hookworm infection is for the most part asymptomatic and, in the view of some auto-immune experts, can provide transformative relief.[16]

The 1970s baby formula debacle highlighted just how important mother's milk is for a developing infant. The natural probiotics that a newborn receives during those first days and months can last a lifetime, and are much healthier than any baby formula manufactured by Nestlé even if filled with vitamins and calories but lacking in microbes. Convenience had been exchanged for profit, with the formula also lacking any antibodies, and made less healthy by pathogen-filled tap waters used to mix it up. The old formulas were eventually taken off the shelf at the same time as companies doubled down on better reproducing a mother's own breastmilk, which could be had unadulterated for free. In similar fashion, time-tested vaginal birth provides the newborn with a dose of probiotics that is not available to the baby who exits the womb via a Caesarian section. Without ever realizing it, many of our daily habits, customs and events have been providing us with efficient venues for transferring useful microbes between parent and offspring, between sibling and sibling, between lovers and between friends, before modernity ever stepped in, and are only now being rediscovered and readopted for the benefits that they can bring.[17]

More aseptic methods of introducing and favouring probiotics may involve *prebiotics*. In this approach to restoring a body's biodiversity, inorganic nutrients and other key growth compounds are brought into the body rather than the beneficial microbes themselves. If healthy creatures already exist in our internal recesses, their numbers and compositions can be enhanced by providing them with optimal substrates.

Microbial deficiencies can thus be addressed, as in the oralome, by chewing specially treated gum. The arginine, galactoside and succinic acid that may be laced into gum arabic can be sloshed back and forth via mastication to create a fertile milieu that helps rejuvenate useful mouth bacteria and healthy biofilms. For resuscitating flora and fauna in the lower intestine, prebiotics in the form of food fibres can be ingested to enhance growth habitat there. Just as a gardener mulches straw or compost into the soil to enrich its microbial numbers and boost fertility, our grandmothers insisted that we ate our roughage for the same purposes. It may come as little surprise that 'roughage' a century ago was primarily a farmer's term, used in reference to soils, and has been progressively co-opted by dieticians eyeing the human gut and pointing out the importance of maintaining healthy microbial growth media.[18]

It turns out that our assumptions about prebiotics have also evolved over the decades. Conceived in the mid-1990s as denoting non-digestible compounds that foster microbial growth, the word 'prebiotics' is coming to mean any material or condition that promotes a rich flourishing of microbiota. Although corporate profit interests maintain that prebiotics are simply physical compounds that can be manufactured and sold, it is clear that such products must also be consumed in the correct proportions, at the right times, under optimal temperatures and rhythms, while interacting with other organisms and aligning with cultural habits and norms. The 10^{14} organisms inhabiting our insides are sensitive to every factor that we hosts subject them to, so that *prebiotics* (as pre-life factors) might really be understood as any material or circumstance that favours more vigorous growth of *probiotics*. A high-fibre meal promotes a richer intestinal microbiome, but for optimal results one can combine it with other prebiotics, such as a balanced diet, eaten in moderate quantities, at the right time of day, in a conducive emotional state while being free from serious disease. Dietary fibre is thus one element of many that helps us maintain a healthy gut biome. The fact that every person fosters his or her own fingerprint of microbiota reflects how many prebiotic factors may be involved in promoting its growth. Our diet, age and gender, our clothing, furniture and workplace, our bathing habits (or lack thereof), our hobbies and our proximity to friends, family and pets all contribute to the unique compositions of creatures living in and on us. Couples share more of their microbiota than do two people living separately. A person and their pet share more of their microbiota than if they lived apart. Subtle prebiotic differences between households can result in vast microbiomal differences. No wonder that your neighbour's home doesn't quite smell like your own.[19]

More recent additions to the conservation microbiologist's lexicon, finally, are 'synbiotics' and 'postbiotics', with the former denoting mixtures of pre- and probiotics, and the latter being anything left over once probiotics die, such as their cell walls, which can continue to promote microbial vitality. Both syn- and postbiotics help restore and enhance bodily biodiversity. The larger point here is that scientists are increasingly convinced that creatures living in or on other creatures are benefiting each other – whether before, during or after they live out their lives – so that they are cooperating with each other at many stages of their existence. It has been a long journey since the days when parasites were assumed to be only troublesome and harmful, in the way that a 1929 textbook defined a parasite to be an organism that 'lives at the expense of another organism that harbors it'. Instead, it seems that many organisms are compensated by the pre-, pro-, syn- and postbiotics they come to harbour. Or, if one insists that parasites are, by definition, harmful to their hosts, then there are far fewer parasites out there than has been previously supposed, and that the remaining *real* parasites would be better labelled 'pathogens'. *Parà•sítos* – παρά•σιτα – eat beside, and most of them are dutifully paying for their meals.[20]

Restoring versus Rewilding Our Microfauna

Bringing back our bodily creatures may seem a preoccupation limited mostly to urban, Western and wealthy folks, as Jamie Lorimer points out in *The Probiotic Planet*. In the example of *Necator americanus*, a common species of hookworm, he observes that a significant proportion of people in the developing world still harbour these helminths in their bodies, and that they are actually rather harmless, even if these people are constantly reminded by healthcare workers of their heavy parasite loads and the potential harms that may result. Of the 500 to 700 million people who carry hookworm, few experience any ill symptoms. As these carriers are administered deworming remedies and convinced to lead more sanitary lifestyles while possibly migrating to cities, they lose large numbers of their microbiota, and begin suffering such postmodern conditions as auto-immune disorders. The final response has been to encourage these people to ingest worms anew. 'Replacement hookworms' are now being reintroduced to their bodies, often imported from the very places that they or their grandparents departed from in a grand cycle of what Lorimer calls 'worming, deworming, and reworming'. The world is now entering a probiotic age in which more and more of us are appreciating and reviving the variety of lifeforms that surround us and inhabit us.[21]

A key issue in this reworming project is deciding on whether we should restore parasite biomes to what they once were, or to improve them beyond the original. As in the larger push to ingest probiotics, reworming may aim to recreate what once circulated in ourselves (and our ancestors), or else create novel kinds and numbers of our internal flora and fauna that can better address our modern needs. Nature conservationists working at the larger scale of wildlife populations and landscapes have been confronting this question for decades, and it is important to realize that parasite conservationists deal with the same issues about whether it is better to restore or else rewild. Restoring depends on a historical record to ascertain how a biota has changed, often by identifying a baseline or standard that represents the ideal target. Rewilding, though, is a more future-oriented pursuit, with conservationists aiming to create optimal levels and kinds of biodiversity based on desire or design rather than on historical conditions. Rewilding can be the same as restoring if the target state is that which existed in a former wild condition – yet it may be impossible to determine the composition of that wild state since nearly every microbiome (like every landscape) has strayed or been modified far beyond any of its primordial reference states. The fact that these references and baselines have shifted means that rewilders need to be flexible in identifying their conservation goals. The project of bringing a mountain meadow back to life after it has been overgrazed, eroded and invaded, for example, is more realistic for a rewilder than a restorer, since this meadow's changing diversity, fertility and hydrology mean that it can never be re-created anew, only converted into a different future state. So, too, in the endeavour of restoring a healthy microbiome, it may be unrealistic to try to bring back a former state, and a much better strategy to rewild one that accounts for current requirements and constraints.[22]

For such reasons, rewilding is increasingly favoured over restoring because of the flexibility that it implies. The project of rewilding Yellowstone, for example, is carried out by reintroducing wolves to a relatively unrestricted landscape, allowing them to reproduce, prey on other animals and reinstate ecological processes in the way that the wolves that once inhabited this place did. Enthusiasts like George Monbiot feel that Yellowstone's free-roaming wolves, as ecosystem engineers and keystone species, are promoting wildness at many levels in the park – even if this ecosystem can never be returned to a facsimile of what it once was. Yet the park can still become a wilder place that can better contend, say, with 4 million human visitors each year. In a similar way, microbiologist Samiran Banerjee explains that the microscopic world is also shaped by keystone organisms that 'exert considerable influence on microbiome structure and functioning'. One can then focus on conserving these microbial keystone species, either by reintroducing them or by promoting existing ones with prebiotics, to foster wildness wherever such species are harboured, as in the lower intestine for improving its ability to contend, say, with novel toxins or new pathogens. If the collective insides of the world's organisms really is 'one of the three major habitats on the earth, comparable to the aquatic and terrestrial habitats on which the hosts themselves dwell', as Warder Allee taught in his classic *Principles of Ecology* (1949), then the project of recreating a healthy microbiome must be guided by the same ecological principles that govern aquatic and terrestrial habitats.[23]

One can also realize that rewilding may produce very different end products depending on how it is practised. In the case of rewilding the Netherland's Oostvaardersplassen landscape reserve, which is sometimes compared with rewilding Yellowstone, the keystone species are grazers rather than carnivores. In Oostvaardersplassen's 56 square kilometres of lowland grasslands near Amsterdam, hardy ancient breeds of pony and cattle – along with artificially bred grazers meant to simulate them – are let loose to begin creating novel landscapes through a process called *naturentwicklung*, or 'natural development'. Unlike Americans who may foster pristine, pre-human conditions in their park, the Dutch encourage spontaneous 'natural development' to guide their reserve, itself won back from the sea through dyking and draining to create what has become a new wild rather than a past nature. Managing our intestinal biodiversity in the Dutch style of *naturentwicklung* may find us consuming designer microbiota for creating a unique gut biome, rather than aiming to reproduce approximations of our primordial microbiota, with the two strategies producing rather different results.[24]

Rewilding therefore has greater potential to meet human needs than does restoring, but it also presents greater risks of sallying forth to produce unknown and untested conditions that are worse than before. One researcher cautions that it is unclear 'where long-term rewilding will lead' since there are many 'uncertain outcomes' of this practice. Even though this researcher has landscapes rather than bodyscapes in mind, rewilding our insides is inherently riskier than restoring them, possibly creating more desirable combinations of microbiota, but also capable of multiplying dangerous, unprecedented microbial regimes. By ingesting loosely defined mixtures of micro-organisms, by receiving faecal transplants from those with very different microbial exposures from us, by being inoculated with varieties of pinworms or roundworms wholly new to our intestines or joining the rising trend of consuming raw, unpasteurized milk – and by generally playing God with our keystone microbiota, we may end up creating a microbiome that leaves us in more serious shape than if we had simply learned to live with an impoverished microbiome. As with conserving external ecosystems, conserving our internal ecosystems may be more wisely carried out by recreating approximations of former baselines than creating wild biomes anew.[25]

Rewilding to the Future

Despite such probiotic warnings, many people continue to rewild their insides even as they inadvertently disrupt and homogenize their microbial world. In modern hospitals today, our bodily biodiversity is simultaneously being impoverished and rewilded in the process of ingesting antibiotics while also contracting unfamiliar contagions, some of which are superbugs invigorated over time by the very chemicals meant to destroy them. The response to this bodily disruption, in ourselves and in our domestic animals, is to generally ingest more probiotics, from viruses to helminths. Despite significant scepticism about prescribing probiotics in farm animals, reports one agricultural journalist, 'there is a continuing stream of research evidence pointing strongly towards their positive role in maintaining a balanced intestinal microflora and lessening the dependance on antimicrobials.' This probiotic turn means, for example, that certain horse breeders now practice 'sustainable deworming' by feeding their horses more roughage and fewer deworming pills. Some dog breeders now blame pet allergies on their animals' impoverished gut flora while buying them niblets saturated with microbiota.[26]

In fact, every day that people go about their business, they are simultaneously domesticating and rewilding microbiomes, within and beyond themselves. Our finest cheeses are seeded with lactic acid cultures, and then transformed with savoury moulds or fungi or passed through the gut of fly larva in time-honoured traditions of controlled rotting. Smelly gold prospectors wandering across the Yukon were nicknamed 'sourdoughs' because of the packets of live yeast dangling around their necks, used piecemeal for leavening flour. We build a septic system and then rewild it by mixing in bacterial supplements for disintegrating the sewage flowing from our toilets. We construct weirs across rivers to help aerate the waters for encouraging microbial growth and improving river health. By incubating a rich microbiota outside of us, we are providing more opportunities for creatures to travel inside of us, in projects that are mixing 'old friends' with novel organisms. Our internal compositions are constantly changing, but we still maintain some ability to modify and shape them.[27]

In the end, wholly new micro-organisms are finding their way into our rewilding designs despite the world trend towards ecological and microbiomal simplification. Darwin taught that species arise through spontaneous mutations followed by natural selection; Margulis that species also originate through symbiogenesis, whereby new organisms emerge out of combinations of existing organisms – as when parasites learn to live peacefully and permanently within their hosts who simultaneously learn to accommodate their parasites. Viruses are found nearly everywhere, even floating down from the atmosphere by the gigatrillions while wafting between continents and always searching for hosts to infect and begin reproducing. Bacteria are also ubiquitous, found not only in clouds but under ocean floors. Just as primordial bacteria apparently entered larger cells for eventually assisting with metabolism as well as photosynthesis, some of today's transient microbes are also being incorporated into new host species to begin cooperating with one another and corroborating Kropotkin's worldview. The biosphere's ubiquitous strands of DNA serve as building blocks for new lifeforms that are becoming part of us, and of the planet. Parasitism, as the condition of host joined with parasite, provides life with new possibilities.[28]

Humanity's Most Unwelcome Guests?

The fantastic voyage into the world of our bodies also reveals that parasitism can be a state of mind. Or, as parasitologist Eric Hoberg sees it, parasites are the key to understanding 'the history of life, the universe, and everything'. Yet beyond borrowing insights from *The Hitchhiker's Guide to the Galaxy*, there really are those who obsess over these creatures – although sometimes with ill effects. As early as the 1920s, *Stedman's Medical Dictionary* was explaining that so many people develop an unrelenting fear of parasites that there is an identifiable condition that can be called parasitophobia whereby, in a typical case, 'the skin itched because tiny bugs which barely could be seen were crawling all over the surface of the skin, in his eyebrows and in his nostrils. He said that the only relief he had was to dig into the skin with his nails and pick out the parasites.' The morbid fear of parasites might manifest itself in various ways, even though few or no real parasites can ever be found. Two decades later, one psychiatrist reviewing his clinical experiences called attention to the same condition, with the additional suggestion that its susceptibility might be linked to gender:

> The patients are mainly women. Age usually 50–60 years. The patients complain of itching and paraesthesia and are stubbornly and fixedly convinced that the complaints are caused by little animals (scabies mites, felt lice, worms and so on). Much time is spent searching for and trying to exterminate the animals.

Also during the general parasitology conference of 1980, Paul Beaver called attention to those stricken by a condition whereby certain individuals aim 'to rid themselves of their parasites even when they are essentially harmless – or even when they are nonexistent ... [We] must expect occasionally to encounter people whose most troublesome parasites are those that exist only *in the mind*.'[29]

The bedbug panic of Paris in 2023 became another manifestation of parasitophobia, in which little creatures were found across hotels and private residences, in linens and towels and mattresses, and scared away tourists. Some Parisians were convinced to begin dumping their belongings, fumigating their apartments and even picking themselves raw, certain as they were that their little bedfellows were gnawing at them at all hours – even when none could be found. Some onlookers blamed the bedbug hysteria on an orchestrated social media campaign, with the French European affairs minister stating that the 'bedbug polemic was in a very large part amplified by accounts linked to the Kremlin, and they even created a false link between the arrival of Ukrainian refugees and the spread of bedbugs'. In the midst of the Rugby World Cup, and anticipating the Summer Olympics a year later, local experts denied that there was any real problem. Of ten bedbug cases reported in the *Metro*, 'all have been checked ... there were zero proven cases'. Of 37 more sightings on local trains, 'all have been checked, zero proven'. But since bedbugs 'affect health, the economy, transport, tourism ... a comprehensive approach' had to be taken, declared another government spokesman. Despite the logic of eyesight, Parisians continued to struggle under the grip of their phobia even though few real bedbugs were ever seen. When New Yorkers faced their own bedbug crisis fifteen years earlier, two members of Congress sponsored a 'Don't Let the Bed Bugs Bite Act', which was designed to bring special resources to exterminating the critters. Government's highest branches were finding themselves caught in the throes of parasitophobia.[30]

And so, despite all that is known about the benefits that parasites can bring to our midst, and the many crucial roles that we now know they play, parasite aversion continues to reach new heights. Most of us still abhor a parasite – and any serious project aimed at restoring some of them would seem doomed. Perhaps our feelings stem from the fact that they are often so small or invisible that we cannot observe them well enough to appreciate all the advantages they provide. Or maybe we have accepted an exaggerated etiquette of cleanliness that does not allow us to welcome our most intimate creatures: finding little crawlers in our scalp or between our toes means that we did not scrub hard enough, or that our undergarments were not washed long enough. More fundamentally, our concept of 'self' is still limited to a single being – ourself – and does not easily account for the billions of co-beings that compose us. Carl Zimmer subtitles his book about parasites as *Nature's Most Dangerous Creatures*. Parasitologist Rosemary Drisdelle titles her own work *Tales of Humanity's Most Unwelcome Guests*. In *The Gospel of Germs*, historian Nancy Tomes tracks how Westerners developed a deep revulsion to their intimate co-travellers, tracing much of that sentiment to the late nineteenth-century germ theory of disease when microbes were assumed to have a proclivity to harm, often being described with 'highly charged adjectives such as "foreign", "base", "murderous", and "cunning"'. Once Koch and Pasteur and their followers convinced us that we really should dread our bodily inhabitants, we have been trying to eradicate them ever since. As physiologist Henry Gradle had articulated a general theory of disease (couched in Darwinian terms in 1883), 'Diseases are to be considered as a struggle between the organism and the parasites invading it . . . and [the struggle] must terminate in the victory of one or the other side.' Even though medical scientists have since clarified that bacteria and other microbes are crucial for digesting our food or synthesizing key vitamins, we only pay attention when we learn that bacteria are also the source of staphylococcal infections and bubonic plague. Seeing our co-creatures as only base and murderous seems unfair, but that is our usual view of them. Walking into the pharmacy or veterinary supply shop finds us stocking up on a few extra bottles of Parasite Cleanse. To finally set our minds at ease, maybe we really should eliminate any proposal to restore parasites or rewild microbiomes, despite the many advantages that our bodies and the rest of the world may reap.[31]

Even the great biodiversity advocate Edward O. Wilson could not help but single out a most-wanted list of creatures, ones that in an ideal world be eliminated from the planet, many of which are parasites: 'The biosphere would not mourn the loss of mosquitoes of the African *Anopheles gambiae* group of species,' wrote Wilson,

> specialists on human blood, expert hiders in native dwellings, and the principal carriers of malaria. Nor, I suspect, would even dedicated conservationists mourn the complete elimination of the African guinea worm . . . Other than still-unknown pathogens among bacteria, microscopic fungi, and viruses, the number of species worthy of extinction, or at least of harmless storage in liquid nitrogen, is probably (my guess) fewer than a thousand.[32]

Undoubtedly reflecting on his own experiences of contracting fevers during fieldwork on distant shores, Wilson felt no remorse in calling for the elimination of the world's principal disease vectors and the creatures they carry. And yet one might expect that of all observers, this renowned naturalist and author of *Biophilia* would point out that even humanity's most troublesome, most infectious and most dangerous organisms can under certain circumstances play vital roles in contributing to the healthy functioning of the Earth's natural systems. In following up on Wilson's nod towards biophobia, a journalist asked for clarification: 'I've talked with other scientists and science journalists who were, frankly, startled to hear that you, E. O. Wilson – the most famous entomologist in the world – say it's OK to eradicate entire mosquito species.' In response, Wilson would only say, 'I always thought we should get rid of *Anopheles gambiae* . . . I'm talking about a very small number of species that have co-evolved with us and are preying on humans, so it would certainly be acceptable to remove them. I believe it's just common sense.'[33]

To such humane and sensible reasoning, one must agree that it is undeniably true that there are some microbes and micro-creatures to be feared, that several are intent on killing and consuming us, and that for a few of them, it is us or them: but these are not centrally parasites but pathogens. Even humanity's most threatening and deadly animal, the mosquito, may occasionally act on our behalf by passing on malaria to strike our enemies harder than ourselves – especially if we properly prepare ourselves with acquired antibodies, G6PD deficiency, negative Duffy antigens and scattered sickle cells. Even the deadly malaria parasite, which is carried by the mosquito and cooperates with it, can also cooperate with us and make life better for both of its hosts. *Anopheles gambiae* armed with parasites, after all, helped keep European invaders out of Nigeria, at least for a time, and is one reason why it might be embroidered on Nigeria's flag, despite Wilson's pronouncements.[34]

Or it may be the *proportion* of parasite and pathogen in each of our bodily creatures that should concern us, with the *Plasmodium* of malaria being part parasite and part pathogen, able to help us or hurt us depending on circumstance. This protozoa that grows in our blood and brings us deep aches and high fevers can, after all, pass on these symptoms more severely to those who threaten us but have never been exposed to malaria. Certainly mosquitoes might also be understood as being largely pathogenic when one simply tallies up malaria's mortality rates around the world, despite the many kinds of mutualism that can exist between this insect and its human host. And even in the case of guinea worms – surely one of the cruellest creatures to infest human beings, wreaking pain and infection upon anyone unlucky enough to drink from a creek teeming with its larvae or else eat a frog or fish already infected by them – even this most vile of creatures may still be considered partially parasitic even if it is overwhelmingly pathogenic. It would be preposterous to think that a guinea worm can ever be an ally to any human being, especially if one finds the head of one poking out of one's swollen foot, which might be carefully tugged and wrapped around a twig for twisting out a few millimetres at a time over a day or two to finally extract the 'little dragon' – or *Dracunculus*. Twist slowly or the worm might snap in two, leaving its lower segment to fester inside the wound. A nineteenth-century encounter in Egypt with the guinea worm offers further gruesome details:

> Frequently, after extracting one worm from a patient, a second, a third, or even a fourth, would appear: after getting one out of a leg, a second would appear in the other, a third in one hand, and a fourth in the other hand. The Guinea-worm, I believe, has been seen in every part of the body.[35]

For the person who suffers such an ordeal, along with the permanent rashes or cysts left over after the guinea worm is gone, there would be little consolation to learn that infections by other helminths, such as pinworms or hookworms, can actually provide significant benefit by heightening a person's immune response. Just as inconsequential for the guinea worm victim would be to learn that other helminth species, such as schistosomes in Senegal or ascaris roundworms in islands near Madagascar, may provide defences against contracting the dreaded malaria itself, as by pitting one bodily inhabitant against another. In fact, it may really be that only someone reading about parasites in the comfort of their living room, or else sifting through the vast literature of parasites in their local air-conditioned library, would ever suggest that a guinea worm is *not* absolutely and always a human enemy. But if there really is 'antagonistic interaction' between many co-infected parasites, as one study suggests, or if public health agencies need to be 'cautious' before implementing widescale control measures of our secondary parasites, as another study claims, then just maybe we can be a little less fearful of guinea worms than we are, for such parasites may be providing us with some defences against even more dangerous organisms. And so there is a small possibility that even guinea worms are not quite as

Surgeons extracting the little dragons, from Georg Hieronymus Welsch, *Exercitatio de vena medinensi, ad mentem Ebnsinae* (1674).

evil as we assume. For all these reasons, I hereby nominate the guinea worm as 99 per cent pathogenic and 1 per cent parasitic – with that 1 per cent keeping open the possibility that even these worms may be acting as discreet mutualists when they *eat beside*.[36]

One may finally ask if the great projects of disease and vector eradication, as when the Rockefeller Foundation oversaw malaria control efforts across South America, Africa, Asia and the Mediterranean during the twentieth century or when the World Health Organization battled COVID-19 across the globe in the twenty-first century, have really been part of humanity's efforts to engineer Gaia for making the

Earth a healthier place. The millions of tons of DDT sprayed across the world, and the billions of mRNA vaccines injected into peoples' arms, would seem to be a welcome rearrangement of ecosystem processes for ushering in greater good for greater numbers, human and non-human alike. But such facile celebration of human disease control omits the fact that there are cooperative dependencies between parasite and host, and that all parasites and even many pathogens are deeply interested in the ultimate survival of their hosts – with hosts standing to lose something if even their mostly pathogenic creatures disappear. Such losses may not be readily apparent or manifested over the short term, but certain malaria-carrying mosquitoes along with the worst parasites that they carry and even the ghastly guinea worm itself – the focus of an intensive control programme now ushering this organism into oblivion – may not be as 'worthy of extinction, or at least of harmless storage in liquid nitrogen' as one assumes. If we cannot figure out ways to restore the benefits provided by our keystone parasites, we may need to consider ways of restoring some of these parasites themselves so that they can continue rewilding ourselves and our world anew.

AFTERWORD:

Parasite Responsibility

At any moment this loathsome parasite may creep into me,
and then – I must tell some one my hideous secret.

ARTHUR CONAN DOYLE, *The Parasite* (1894)

S
ome years ago, before ever sitting down to write this book, I was
at my desk with sandwich in hand clicking through my email list,
and then spotted the subject heading 'Response to your Sardinia
Paper' – sent by someone whom I did not recognize. Since I had written
some early pieces about Sardinia's experiences with malaria, here was
a message that was probably not junk mail and merited actual opening
and reading to learn what my correspondent had to say.

'I came across your paper and found it really startling,' began the
note.

> My father is Sardinian – he is one of 11 children. He is the second
> youngest, born in 1932. His eldest sister must be about 100yrs
> old now. His eldest brother died of something like bone/blood
> cancer when he was about 50yrs old and another brother died
> of bone marrow cancer when he was about 55yrs old. 2 of his
> sisters had some kind of blood disorders that they needed to
> have blood transfusions for and they have since passed away in
> the last couple of years. Another of his brothers lost his eyesight
> because they gave him Quinine during the war.

Since I am not used to receiving family health histories in my personal email, I didn't know whether to delete and continue with my sandwich or keep reading. It was really none of my business. But I was curious:

> I have had health issues lately and I went to see a Naturopath; he asked me about 3 times if I had ever been around DDT or heavy metals. On the Hemaview screening report it picked up on Grade 4 Acanthocytes – the report says 'The spiny projections on these red blood cells may be caused by a number of metabolic changes in the body, including possible defects in the liver function or an imbalance in fat metabolism'. I have gynecological problems at the moment and the gynea sent me for cancer marker tests – they came back yesterday and they were all ok.

At this point, I felt that I could not offer much in response to this note, deciding that this was a message that I need not, and probably should not, answer. We university types have classes to teach and articles to write and we strive to be sympathetic, but not nosy. The message continued:

> I was just searching about the Sardinian diet on the internet (because I want to eat more healthily) and came across articles about malaria and then stumbled upon your paper. Check out this article from *Time* magazine on how long some people live for in Sardinia ... 'The traditional cuisine of Sardinia is in some ways a contradiction: An island civilization that did not utilize seafood in its diet. Since Sardinia's coast has always been victim to invasion, the Sardinian people found refuge in the mountainous hinterland of the island. Therefore the traditional foods of Sardinia were always more influenced by the land than the sea. Gets one thinking ...
>
> Kind regards, Lara (Perth, Australia)

I closed the email and then clicked through the rest of my inbox, deleting most messages while addressing a few others. But I let this immediate one sit.

Hours later, before departing for home that day, I decided that here was a message that deserved an answer. I did know a few things about the health of Sardinians and their island, after all, especially about some of the co-evolutionary diseases that are common in malarial (and ex-malarial) places – and I had come to suspect that my unsolicited correspondent might actually be suffering from certain health issues linked to her Sardinian heritage. Of course, I wasn't going to try to diagnose her problems from afar, but I realized it was a sort of obligation to mention some of the insights that I had learned during my research. So in offering a quick response to her note, I provided a link to a study which concluded that repeated exposure to DDT, which had been heavily sprayed across Sardinia during the late 1940s for controlling malaria-carrying mosquitoes, should not present very serious human health risks, particularly across generations – since this study concluded that most DDT sprayers associated with that project did not later suffer from health problems that were out of the ordinary with those of the rest of population. After jotting down a few other odds and ends, I mentioned that she is probably endowed with very good genes for longevity stemming from her Sardinian heritage, and then quipped, 'I could suggest lots of olive oil, and plenty of beach sunshine, just like they get in Sardinia.' Before sending off my note and shutting down my computer, I added a postscript:

> Oh, I should also mention that Sards do tend to suffer from favism at higher rates, an unusual condition whereby one cannot digest fava (and related) beans very well; it has to do with genetic *adaptations* people have developed in highly malarial places ... sort of like the famous sickle-cell anemia that some people from ex-malarial areas confront. You might want to be on the alert for favism and its symptoms. I knew I should have been an MD, but got this PhD instead.

Following two weeks of silence, and having all but forgotten the candid message I sent out that afternoon, I received a follow-up note from my Australian contact:

> Dear Marcus: Thank you so much for giving me the clues I was missing – you are an angel. It looks like almost everyone in my father's Sardinian family carries the gene for G6PD deficiency. I have had all the symptoms. I ate a whole big plate of chick peas on the Thursday night – I was trying to be healthy. I received your email on the Friday morning. I was in hospital emergency on Friday after with tachycardia, palpitations, pallor, slight fever, raised blood pressure. I ended up again at hospital emergency on Wednesday with the same symptoms.
>
> The doctors are all stumped even though I have given them the answer. The only doctor who kind of believed me was Chinese whose sister has G6PD deficiency. But he still didn't think I have it.

My correspondent went on to list several weblinks to articles detailing various studies about health conditions, including a few about the intricacies of G6PD deficiency – or favism – before signing off: 'Take care & God Bless. Kind regards, Lara'.

Here at long last, I realized with this brief email exchange, was evidence for 'the uses of history' – and in this case, the uses of health history or perhaps, of parasitic history – since malaria stems from a microscopic parasite. Having studied health and environmental histories for a good long time, I have more than once confronted the tough question about the utility of such research. The stories that historians tell, after all, may be interesting, and some of them may even help us ponder whether 'those who do not learn history are doomed to repeat it'. But can the historical craft offer actual remedies to our immediate, pressing challenges? The present book has kept that question in mind. I am now more convinced than ever that our most intimate bodily creatures need to be placed in larger historical context: such organisms and our time with them provide us with completed and ongoing experiences that can lead to useful insights. The foregoing chapters may be considered post-postscripts to Lara's note. It is the union of ourselves and our fellow travellers that have such wide implications for all of us. Take care and thank you, Lara.

One must therefore admire, but also question, the conclusions that our forerunners taught us. More than three hundred years ago, the English astrologer and physician William Ramesey described a vivid encounter with our bodily creatures, noting in his remarkable *Theologico-Philosophical Dissertation concerning Worms in all Parts of Human Bodies*, that 'if our Eye-sight were enlarg'd, or our Bodies set in a true Light, we should appear to be the most amazing Spectacle in the whole World':

> There should we see an infinite Number of Worms swimming in the Blood, and sallying from the Heart thro' the Arteries, and returning back by the Veins: There should we see Thousands of living Animals of various Shapes and Sizes, crawling in the Eyes, Nose, and Ears; the very Mouth fill'd with them, the Tongue stuff'd full of them, the Gums tormented, and the Teeth excavated by them. Nay, we should see not only the Brain full of them, but the Flesh abounding with them, and the very Bones perforated by them; and Thousands every Moment crawling thro' the Pores of the Skin.

Ramesey's main purpose in his lucid descriptions was to offer the warning that these 'vast Swarms' of animals and 'Mass of Worms' were continually feeding on us, so that in the end such creatures would 'gradually destroy' us. But far from destroying us, the vast majority of creatures living in and on us are also helping us to live healthier, even longer lives. Despite Ramesey's pronouncements, we can realize that we live in a mutualistic world in which parasites and hosts must act responsibly towards one another.

REFERENCES

PREFACE: A Wormy World

1 For estimates of parasite numbers that inhabit humans around the world, see Richard Ashford and William Crewe, *The Parasites of Homo Sapiens: An Annotated Checklist of the Protozoa, Helminths and Arthropods for Which We Are Home* (London, 2003); Michael Balter, 'Taking Stock of the Human Microbiome and Disease', *Science*, XXXIII/6086 (2018), pp. 1246–7; 'The Human Microbiome Project', PLOS Partnered Collections, Special Issues (2018); see also Curezone, 'Human Parasites', http://curezone.com/ diseases/parasites; 'Pinworm (Enterobiasis)', www.humanillnesses.com; 'The Human Microbiome Project', https://collections.plos.org, 24 May 2018; Shyamala Iyer, 'Building Blocks of Life', Arizona State University, http://askabiologist.asu.edu, all accessed 14 November 2024.
2 William Morton Wheeler, *Foibles of Insects and Men* (New York, 1928), p. 52.
3 Noreen Lacey, Síona Ní Raghallaigh and Frank Powell, 'Demodex Mites: Commensals, Parasites or Mutualistic Organisms?', *Dermatology*, CCXXII/2 (2011), pp. 128–30; Martin Blaser, *Missing Microbes: How the Overuse of Antibiotics Is Fueling Our Modern Plagues* (New York, 2014).
4 James Lovelock, 'Beware: Gaia May Destroy Humans Before We Destroy the Earth', *The Guardian*, www.theguardian.com, 2 November 2021.
5 P. Zaccone et al., 'Parasitic Worms and Inflammatory Diseases', *Parasite Immunology*, XXVIII/10 (2006), pp. 515–23.

1 Sardinians Do It Longer

1 D. H. Lawrence, *Sea and Sardinia* (New York, 1921), pp. 159–61.
2 Ibid., p. 102; L. L. Cavalli-Sforza, P. Menozzi and A. Piazza, *The History and Geography of Human Genes* (Princeton, NJ, 1994), p. 268; Tracey Heatherington, *Wild Sardinia: Indigeneity and the Global Dreamtimes of Environmentalism* (Seattle, WA, 2010); M. Sikora et al., 'Population Genomic Analysis of Ancient and Modern Genomes Yields New Insights into the Genetic Ancestry of the Tyrolean Iceman and the Genetic Structure of Europe', PLOS *Genetics* (May 2014), doi: 10.1371/journal.pgen.
3 G. Caselli and D. Rasulo, 'Centenarians in Sardinia: The Underlying Causes of the Low Sex Ratio', paper delivered at meeting of the International Union for the Scientific Study of Population, Tours, 2005, http://iussp2005.princeton.edu,

accessed 7 March 2012; J. M. Robine et al., 'An Unprecedented Increase in the Number of Centenarians', International Union for the Scientific Study of Population, www.iussp.org, accessed 7 March 2012; Jason Wilson, 'How to Live Forever', *The Smart Set*, www.thesmartset.com, 6 August 2007; Nick Squires, 'Sardinian Siblings Aged 818 Officially the World's Oldest', *The Telegraph*, www.telegraph.co.uk, 21 August 2012; U.S. Census Bureau, 'The Older Population: 2010', www.census.gov, accessed 7 March 2012, p. 18; UK Office of National Statistics, 'Census 2021', www.ons.gov.uk, accessed 25 March 2024.

4 Dan Buettner, *The Blue Zones: Lessons for Living Longer from the People Who've Lived the Longest* (Washington, DC, 2008), pp. 35–6.

5 Lawrence, *Sea and Sardinia*, p. 157.

6 William Henry Smyth, *Sketch of the Present State of the Island of Sardinia* (London, 1828), p. 168, quoted in Buettner, *Blue Zones*, p. 36.

7 Buettner, *Blue Zones*, pp. 62–3.

8 M. Poulain et al., 'Identification of a Geographic Area Characterized by Extreme Longevity in the Sardinia Island: The AKEA Study', *Experimental Gerontology*, XXXIX/9 (2004), pp. 1423–9.

9 Maurice le Lannou, *Pâtres et paysans de la Sardaigne* (Tours, 1941), p. 73; Sir Macfarlane Burnet, *Natural History of Infectious Disease* [1940] (Cambridge, 1972), p. 232; Eugenia Tognotti, 'Malaria in Sardinia', *International Journal of Anthropology*, XIII/3–4 (1998), pp. 237–42; Peter J. Brown, 'Microparasites and Macroparasites', *Cultural Anthropology*, II/1 (1987), p. 162; Margaret Humphreys, *Malaria: Poverty, Race, and Public Health in the United States* (Baltimore, MD, 2001), p. 94; H. F. Gray and R. E. Fontaine, 'A History of Malaria in California', *Proceedings and Papers of the Twenty-Fifth Annual Conference of the California Mosquito Control Association, Inc.* (San Jose, CA, 1957), pp. 18–37; Rosemary Brunetti, 'Outbreak of Malaria with Prolonged Incubation Period in California, a Nonendemic Area', *Science*, 119 (1954), pp. 74–5.

10 Another representative map of malaria showing dark tones in Sardinia is 'Carta della Malaria dell'Italia' by Luigi Torelli (Florence, 1882); Lewis Hackett, *Malaria in Europe: An Ecological Study* (London, 1937), p. 17; Marcus Hall, 'Environmental Imperialism in Sardinia: Pesticides and Politics in the Struggle against Malaria', in *Nature and History in Modern Italy*, ed. Marco Armiero and Marcus Hall (Athens, OH, 2010), pp. 70–86.

11 'Malaria: La Sardegna ricorda i 60 anni dalla "liberazione". Attenzione però ai flussi migratori', *Sardegna Ventirighe*, 28 September 2010.

12 Qiao Liu et al., 'Trends of the Global, Regional and National Incidence of Malaria in 204 Countries from 1990 to 2019 and Implications for Malaria Prevention', *Journal of Travel Medicine*, XXVIII/5 (2021), pp. 1–10.

13 WHO, 'World Malaria Report 2019', www.who.int, accessed 13 April 2020; Frank Livingstone, 'Malaria and Human Polymorphisms', *Annual Review of Genetics*, 5 (1971), pp. 33–64. Another estimate places 2010 malaria deaths at 1.2 million; see C. Murray et al., 'Global Malaria Mortality between 1980 and 2010: A Systematic Analysis', *The Lancet*, CCCLXXVII/9814 (2012), pp. 413–31. According to the Population Reference Bureau, the total number of humans

ever born is some 109 billion, and some 4 to 5 per cent of them, says Brian Faragher, have probably died from malaria; see Carl Haub, 'How Many People Have Ever Lived on Earth?', www.prb.org, 15 November 2022; Ross Pomeroy, 'Has Malaria Really Killed Half of Everyone Who Ever Lived?', www.realclearscience.com, 3 October 2019.

14 Wataru Iijima, 'Colonial Medicine and Malaria Eradication in Okinawa in the Twentieth Century: From the Colonial Model to the United States Model', in *Disease, Colonialism, and the State: Malaria in Modern East Asian History*, ed. Ka-che Yip (Hong Kong, 2009), pp. 61–70; L. Rosero-Bixby, 'Evaluación del impacto de la reforma del sector salud en Costa Rica', *Revista Panamericana de salud Pública*, 15 (1984), pp. 94–103. For illustrative examples of spurious correlations, see Tyler Vigen's book and website that show that there is amazing overlap of the marriage rate in Kentucky and the number of people drowning from a fishing boat (r=0.952), as well as the link between the number of non-commercial space launches and the number of sociology doctorates awarded in the u.s. (r=.789), www.tylervigen.com, accessed 6 June 2015.

15 Sebastian Brandhorst et al., 'Fasting-Mimicking Diet Causes Hepatic and Blood Markers Changes Indicating Reduced Biological Age and Disease Risk', *Nature Communications*, xv/1309 (2024), doi: 10.1038/s41467-024-25260-9.

16 who, 'Malaria', Fact Sheet no. 94 (2012), www.who.int, accessed 12 June 2012. For an overview of malaria in world history, see James L. A. Webb Jr, *Humanity's Burden: A Global History of Malaria* (Cambridge, 2009). For an overview of mosquito control techniques and their challenges, see Marcus Hall and Dan Tamïr, eds, *Mosquitopia: The Place of Pests in a Healthy World* (London, 2021). It is generally agreed that there are five main *Plasmodium* species that can infect humans: *Plasmodium vivax, Plasmodium falciparum, Plasmodium ovale, Plasmodium malariae, Plasmodium knowlesi*.

17 who, 'Fact Sheets: Malaria', www.who.int, 11 December 2024; Joseph Vinetz and Robert Gilman, 'Editorial: Asymptomatic Plasmodium Parasitemia and the Ecology of Malaria Transmission', *American Journal of Tropical Medicine and Hygiene*, lxvi/6 (2002), p. 639; Gillian Stresman et al., 'A Method of Active Case Detection to Target Reservoirs of Asymptomatic Malaria and Gametocyte Carries in a Rural Area in Southern Province, Zambia', *Malaria Journal*, ix/265 (2010), doi: 10.1186/1475-2875-9-265; Denise Doolan, Carlota Dobaña and Kevin Baird, 'Acquired Immunity to Malaria', *Clinical Microbiology Reviews*, 22 (2009), pp. 13–36; C. Onyenekwe et al., 'Prevalence of Asymptomatic Malaria Parasitaemia amongst Pregnant Women', *Indian Journal of Malariology*, xxxix/3–4 (2002), pp. 60–65.

18 N. Bailey, *The Biomathematics of Malaria* (London, 1982); Thomas Wellems et al., 'The Impact of Malaria Parasitism: From Corpuscles to Communities', *Journal of Clinical Investigation*, cxix/9 (2009), pp. 2496–505; Simon Chang et al., 'Long-Term Effects of Early Childhood Malaria Exposure on Education and Health: Evidence from Colonial Taiwan', Discussion Paper Series, no. 5526, Institute for the Study of Labor (iza), Bonn, Germany, February 2011.

19 D. P. Kwiatkowski, 'How Malaria Has Affected the Human Genome and
    What Human Genetics Can Teach Us about Malaria', *American Journal of
    Human Genetics*, LXXVII/2 (2005), pp. 171–92; P. W. Hedrick, 'Population
    Genetics of Malaria Resistance in Humans', *Heredity*, CVII/4 (2011),
    pp. 283–304; Steve Lindsay, Paul Emerson and Derek Charlwood,
    'Reducing Malaria by Mosquito-Proofing Houses', *Trends in Parasitology*,
    XVIII/11 (2002), pp. 510–14.
20 Saheli Sadanand, 'Malaria: An Evaluation of the Current State of Research
    on Pathogenesis and Antimalarial Drugs', *Yale Journal of Biological Medicine*,
    LXXXIII/4 (2010), pp. 185–91; Marc Lipsitch, Martin Nowak and Edward
    Herre, 'Host Population Structure and the Evolution of Virulence:
    A "Law of Diminishing Returns"', *Evolution*, XLIX/4 (1996), pp. 743–8.
21 A. Fortin et al., 'Susceptibility to Malaria as a Complex Trait: Big Pressure
    from a Tiny Creature', *Human Molecular Genetics*, XI/20 (2002), pp. 2469–78;
    Peter J. Brown, 'Cultural Adaptations to Endemic Malaria in Sardinia',
    Medical Anthropology, V/3 (1981), pp. 313–39; Gary Paul Nabhan, *Why
    Some Like It Hot: Food, Genes, and Cultural Diversity* (Covelo, CA, 2006),
    pp. 63–91.
22 Paget Davies, 'Favism', *Postgraduate Medical Journal*, XXXVII/430 (August
    1961), pp. 477–80; Sergio Brau et al., 'La Fava: coltivazione e ricette', ERSAT
    Servizio Circondariale di Cagliari (1997), p. 10; Peter Brown, 'Cultural and
    Genetic Adaptations to Malaria: Problems of Comparison', *Human Ecology*,
    XIV/3 (1986), pp. 311–32; Rosalind E. Howes et al., 'G6PD Deficiency
    Prevalence and Estimates of Affected Populations in Malaria Endemic
    Countries: A Geostatistical Model-Based Map', PLOS *Medicine*, IX/11 (2012),
    p. 9:e1001339.
23 Brown, 'Cultural and Genetic Adaptations to Malaria: Problems of
    Comparison', pp. 311–32; Heatherington, *Wild Sardinia*, p. 90; Lawrence,
    Sea and Sardinia, pp. 182–3.
24 Marcus Hall, 'Thinking Like a Parasite', in *Landscapes, Natures, Ecologies:
    Italy and the Environmental Humanities*, ed. S. Iovino, E. Cesaretti and
    E. Past (Charlottesville, VA, 2018), pp. 117–28.
25 Marcus Hall, 'Today Sardinia, Tomorrow the World: Killing Mosquitoes',
    BardPolitik: The Bard Journal of Global Affairs, V (2004), pp. 21–8; Marcus
    Hall, 'World War II and the Axis of Disease', in *War and the Environment:
    Military Destruction In the Modern Age,* ed. Charles Closmann (College
    Station, TX, 2009), pp. 112–31; Hall, 'Environmental Imperialism in Sardinia',
    pp. 80–81.
26 Christina Frank et al., 'Epidemic Profile of Shiga-Toxin-Producing
    Escherichia coli O104:H4 Outbreak in Germany', *New England Journal of
    Medicine*, CCCLXV/19 (2011), pp. 1771–80.
27 Matt Kaplan, 'Parasites Suck Toxins from Sharks', *Nature*, 25 June 2007;
    M. Malek et al., 'Parasites as Heavy Metal Bioindicators in the Shark
    Carcharhinus dussumieri from the Persian Gulf', *Parasitology*, CXXXIV/7
    (2007), pp. 1053–6; R. van Ommeren and T. Whitham, 'Changes in
    Interactions Between Juniper and Mistletoe Mediated by Shared Avian
    Frugivores: Parasitism to Potential Mutualism', *Oecologia*, CXXX/2
    (2002), pp. 281–8; David Lincicome, 'The Goodness of Parasitism:

A New Hypothesis', in *Aspects of the Biology of Symbiosis*, ed. Thomas Cheng (Baltimore, MD, 1971), pp. 139–227.

28 Charles Elton, *Animal Ecology* (New York, 1927), p. 75; Ryan Hechinger et al., 'Parasites', in *Metabolic Ecology: A Scaling Approach*, ed. R. M. Sibly, J. H. Brown and A. Kodric-Brown (Chichester, 2012), p. 235.

29 Andy Dobson et al., 'Homage to Linnaeus: How Many Parasites? How Many Hosts?', PNAS, CV/Suppl 1 (August 2008), pp. 11482–9; Robert Dunn et al., 'The Sixth Coextinction: Are Most Endangered Species Parasites and Mutualists?', *Proceedings of the Royal Society B*, XXLXXVI/1670 (2009), pp. 3037–45; Andreas Wagner, *Paradoxical Life: Meaning, Matter, and the Power of Human Choice* (New Haven, CT, 2011), p. 163; M. Vannier-Santos and H. Lenzi, 'Parasites or Cohabitants: Cruel Omnipresent Usurpers or Creative "Éminences Grises"', *Journal of Parasitology Research* (January 2011), doi: 10.1155/2011/214174.

30 Malek et al., 'Parasites as Heavy Metal Bioindicators in the Shark', pp. 1053–6; Joseph Jackson et al., 'Immunomodulatory Parasites and Toll-Like Receptor-Mediated Tumor Necrosis Factor Alpha Responsiveness in Wild Mammals', *BMC Biology*, VII/16 (2009), doi: 10.1186/1741-7007-7-16.

2 Adventures in Parasitology

1 E. P. Hoberg et al., '*Umingmaxksfrongylus pallikuukensis* gen.nov. et sp.nov. (Nematoda: Protostrongylidae) from Muskoxen, *Ovibos moschafus*, in the Central Canadian Arctic, with Comments on Biology and Biogeography', *Canadian Journal of Zoology*, LXXIII/12 (1995), pp. 2266–82.

2 Henry B. Ward, 'The Parasitic Worms of Man and the Domestic Animals', *Studies from the Zoological Laboratory*, University of Nebraska-Lincoln, Paper 9 (1895), p. 229, http://digitalcommons.unl.edu/zoolabstud/9, accessed 16 May 2025.

3 P.C.C. Garnham, *Progress in Parasitology* (London, 1971), pp. 131–49.

4 Kenneth S. Warren and Eli Chernin quoted in 'Transcripts of 1980 Conference; Current Status and Future of Parasitology', in Box E, Folder 'History of Definition of Parasitology', John S. Andrews Papers, Special Collections, U.S. National Agricultural Library.

5 Paul C. Beaver quoted in 'Transcripts of 1980 Conference; Current Status and Future of Parasitology'. Most but not all of these transcripts are printed in Kenneth Warren and Elizabeth Purcell, eds, *The Current Status and Future of Parasitology* (New York, 1981).

6 ARTFL Encyclopédie Project, Robert Morrissey, General Editor, 'Parasite', http://encyclopedie.uchicago.edu, accessed 3 March 2013; Friedrich Schiller, *The Parasite; or, The Art to Make One's Fortune. A Comedy in Five Acts*, trans. J.S.S. Rothwell (Munich, 1859); Anders Gullestad, 'Parasite', *Political Concepts: A Critical Lexicon*, www.politicalconcepts.org, accessed 19 July 2013.

7 Victorian zoologist Ray Lankester taught that parasites had degenerated from free-living forms. See Carl Zimmer, *Parasite Rex: Inside the Bizarre World of Nature's Most Dangerous Creatures* (New York, 2000), p. 15; Pierre Joseph van Beneden, *Animal Parasites and Messmates*, 2nd edn (London, 1876), p. 1.

8 Michael Worboys, 'The Emergence and Early Development of Parasitology',
   in *Parasitology: A Global Perspective*, ed. Kenneth Warren and John Bowers
   (New York, 1983), pp. 1–18; John Farley, 'Parasites and the Germ Theory
   of Disease', in *Framing Disease: Studies in Cultural History*, ed. Charles
   Rosenberg and Janet Golden (New Brunswick, NJ, 1992), p. 34; Jean
   Théodoridès, 'Les grandes étapes de la parasitologie', *Clio Medica*, I (1966),
   pp. 185–208; John Farley, 'Parasites and the Germ Theory of Disease',
   Milbank Quarterly, LXVII/Suppl 1 (1989), pp. 50–68; Ernesto Capanna,
   'Grassi versus Ross: Who Solved the Riddle of Malaria?', *International
   Microbiology*, IX/1 (2006), pp. 69–74; Robert Desowitz, *Malaria Capers:
   Tales of Parasites and People* (New York, 1991), pp. 195–8.
9 David I. Grove, *A History of Human Helminthology* (Wallingford, 1990), p. 13;
   Worboys, 'Emergence and Early Development of Parasitology', pp. 1–18.
10 Maurice C. Hall, 'Is Parasitology a Science?', *Journal of Parasitology*, XIX/3
   (1933), pp. 183–91.
11 Benjamin Schwartz and Paul Harwood, 'Maurice Crowther Hall as a
   Parasitologist', *Journal of Parasitology*, XXIV/4 (1938), p. 287.
12 Maurice Hall, quoted in Susan Jones, *Valuing Animals: Veterinarians and
   Their Patients in Modern America* (Baltimore, MD, 2003), p. 126; Maurice
   C. Hall and Winthrop D. Foster, 'Oil of Chenopodium and Chloroform
   as Anthelmintics', *Journal of the American Medical Association*, LXIX/24 (1917),
   pp. 1961–3; Megan McCarthy, 'Chenopodium ambrosioides', https://bioweb.
   uwlax.edu/bio203/2011/mccarthy_mega, accessed 19 December 2024.
13 Willard Becklund, 'The National Parasite Collection at the Beltsville
   Parasitological Laboratory', *Journal of Parasitology*, LV/2 (1969), pp. 375–80;
   Maurice C. Hall, 'Some Practical Principles of Anthelmintic Medication',
   Journal of Parasitology, XIII/1 (1926), pp. 16–24.
14 John S. Andrews, in Box E, Folder 'History of Definition of Parasitology',
   in Box B, Folder 'First 40 Years by Dr. John Andrews' (n.d.), in John S.
   Andrews Papers, Special Collections, U.S. National Agricultural Library;
   Gary Lake, 'Carbon Tetrachloride: A Drug Proposed for the Removal of
   Hookworms, with Special Reference to Its Toxicity for Monkeys When
   Given by Stomach Tube in Repeated Doses', *Public Health Reports*, 17
   (1922), p. 1123.
15 Paul C. Beaver quoted in 'Transcripts of 1980 Conference; Current Status
   and Future of Parasitology'; Warren and Purcell, eds, *The Current Status and
   Future of Parasitology*, p. 4; John Ettling, *The Germ of Laziness: Rockefeller
   Philanthropy and Public Health in the New South* (Cambridge, MA, 1981),
   pp. 209–12; Centers for Disease Prevention and Control, 'Parasites –
   Hookworm', www.cdc.gov, accessed 15 July 2013; Simon Booker et al.,
   'Human Hookworm Infection in the 21st Century', *Advances in Parasitology*,
   LVIII (2004), pp. 197–288.
16 Ettling, *The Germ of Laziness*, pp. 166–7; B. F. Kaup, *Animal Parasites and
   Parasitic Disease* [1914] (Chicago, IL, 1917), p. 9; Maurice C. Hall, *Control
   of Animal Parasites: General Principles and Their Application* (Evanston, IL,
   1936), pp. 8–19. For more about the military metaphors of battling pests,
   see Edmund Russell, *War and Nature: Fighting Humans and Insects with
   Chemicals from World War I to Silent Spring* (New York, 2001).

17 Van Beneden, *Animal Parasites and Messmates*, pp. 2, 85.

18 Jean Baer, *Ecology of Animal Parasites* (Urbana, IL, 1951), p. 6; van Beneden, *Animal Parasites and Messmates*, pp. xxiv, 72–82.

19 Jan Sapp, *Evolution by Association: A History of Symbiosis* (New York, 1994), p. 18; van Beneden, *Animal Parasites and Messmates*, p. xviii.

20 Daniel Todes, *Darwin without Malthus: The Struggle for Existence in Russian Evolutionary Thought* (New York, 1989), pp. 126–30; Charles Darwin, *On the Origin of Species* [1859] (London, 1861), pp. 79, 225; Peter Kropotkin, *Mutual Aid: A Factor of Evolution* [1902] (New York, 2009), pp. 2, 10.

21 Karl Kessler (1879) quoted in Todes, *Darwin Without Malthus*, p. 112.

22 Michael Osborne, 'Parasitology, Zoology, and Society in France, ca. 1880–1920', in *Biological Individuality: Integrating Scientific, Philosophical, and Historical Perspectives*, ed. Scott Lidgard and Lynn K. Nyhart (Chicago, IL, 2017), pp. 206–24.

23 Albert Frank (1877) quoted in Sapp, *Evolution by Association*, pp. 6, 12–14.

24 Kropotkin, *Mutual Aid*, p. xvii; Peter Kropotkin, *The Conquest of Bread* [1892] (London, 1906), p. 12; Eugene Odum, *Fundamentals of Ecology* [1953], 3rd edn (Philadelphia, PA, 1971), p. 220; Peter Kropotkin, *Il mutuo appoggio: fattore dell'evoluzione* [1925] (Bologna, 1950).

25 Sapp, *Evolution by Association*, p. 18; Hermann Reinheimer, *Evolution by Co-Operation: A Study in Bio-Economics* (London, 1913), pp. 14, 90; Hermann Reinheimer, *Symbiogenesis: The Universal Law of Progressive Evolution* (London, 1915), p. xiii.

26 Paul Portier, *Les Symbiotes* (Paris, 1918); Sapp, *Evolution by Association*, pp. 51, 77; Walter Koenig et al., 'Effects of Mistletoe [*Phoradendron villosum*] on California Oaks', *Biology Letters*, XIV/6 (2018), doi: 10.1098/rsbl.2018.0240.

27 Reinheimer, *Symbiogenesis*, pp. 67, 155.

28 George Nuttall, 'Symbiosis in Animals and Plants', Address to Physiology Section (1923), *British Association for the Advancement of Science: Report* (London, 1924), p. 213; Theobald Smith, *Parasitism and Disease* (Princeton, NJ, 1934), pp. 18, 160; Pierre-Olivier Méthot, 'Why Do Parasites Harm Their Host? On the Origin and Legacy of Theobald Smith's "Law of Declining Virulence"', *History and Philosophy of the Life Sciences*, XXXIV/4 (2012), pp. 561–601.

29 E. M. Freeman, 'In Praise of Parasitism', *Scientific Monthly*, XLIV/1 (1937), pp. 67–76.

30 Reinheimer, *Symbiogenesis*, pp. 11, 35; Smith, *Parasitism and Disease*, pp. 44, 48; Theodosius Dobzhansky (1951) quoted in Sapp, *Evolution by Association*, p. 156; Joshua Lederberg (1993) quoted in Méthot, 'Why do Parasites Harm Their Host?', p. 567.

31 Joseph Schall, 'Parasite Virulence', in *The Behavioural Ecology of Parasites*, ed. E. E. Lewis, J. F. Campbell and M.V.K. Sukhdeo (Oxford, 2002), pp. 283–313.

32 Friedrich Küchenmeister (1855) quoted in Grove, *A History of Human Helminthology*, p. 392.

33 L. J.-B. Bérenger-Féraud, 'Distribution geographique des ténias de l'homme', *Bulletin Academy de Medicine de Paris*, Au. 56, 3. s., XXVIII/33 (August 1892), pp. 282–304; Grove, *A History of Human Helminthology*, pp. 392, 405, 406, 409.

34 Warder C. Allee (1929) quoted in Gregg Mitman, *The State of Nature: Ecology, Community, and American Social Thought, 1900–1950* (Chicago, IL, 1992), p. 78; Warder C. Allee et al., *Principles of Animal Ecology* (Philadelphia, PA, 1949), p. 257; Lee Alan Dugatkin, *Cooperation among Animals: An Evolutionary Perspective* (New York, 1997), p. 8; Radhakamal Mukerjee, *The Regional Balance of Man: An Ecological Theory of Population*, Sir William Meyer Foundation Lectures, 1935–2 (Madras, 1938), p. 313.

35 Beaver quoted in 'Transcripts of 1980 Conference; Current Status and Future of Parasitology'. For an overview of evolutionary fitness of cooperation, see Dugatkin, *Cooperation among Animals*.

3 The Benefits of Being Parasitized

1 Gary Fry and John Moore, '*Enterobius vermicularis*: 10,000-Year-Old Human Infection', *Science*, CLXVI/3913 (1969), p. 166; Vaughn Bryant and Glenna Dean, 'Archaeological Coprolite Science: The Legacy of Eric O. Callen (1912–1970)', *Palaeogeography, Palaeoclimatology, Palaeoecology*, CCXXXVII/1 (2006), pp. 51–6.

2 David I. Grove, *A History of Human Helminthology* (Wallingford, 1990), p. 447; 'Digestive Health Centre', www.webmd.boots.com, accessed 16 March 2015; A. Araujo and L. Ferreira, '[Oxyuriasis and prehistoric migrations]', *Historia, Ciencias, Saude-Manguinhos*, II/2 (1995), pp. 99–109; 'Structure, Function, and Diversity of the Healthy Human Microbiome: The Human Microbiome Project Consortium', *Nature*, CDLXXXVI (2012), pp. 207–14; R. Tito et al., 'Insights from Characterizing Extinct Human Gut Microbiomes', *PLOS ONE*, VII/12 (2012), p. e51146, p. 3; Jack Gilbert, Janet Jansson and Rob Knight, 'The Earth Microbiome Project: Successes and Aspirations', *BMC Biology*, XII/69 (2014), doi: 10.1186/s12915-014-0069-1.

3 Klaus Oeggl, 'The Significance of the Tyrolean Iceman for the Archaeobotany of Central Europe', *Vegetation History and Archaeobotany*, XVIII/1 (2009), pp. 1–11; H. Aspöck, H. Auer, O. Icher and W. Plazer, 'Parasitological Examination of the Iceman', in *The Iceman and his Natural Environment*, ed. S. Bortenschlager and K. Oeggle (Vienna, 2000), pp. 127–36; 'Ötzi's First-Aid Kit', www.iceman.it/en/node/288, accessed 19 March 2015; 'USAID's Neglected Tropical Diseases Program', www. neglecteddiseases.gov, accessed 19 March 2015; Tito et al., 'Insights from Characterizing Extinct Human Gut Microbiomes'; Frank Maixner et al., 'The 5300-Year-Old Helicobacter Pylori Genome of the Iceman', *Science*, CDLXXVI (January 2016), pp. 162–5; Martin Blaser, 'Stop the Killing of Beneficial Bacteria', *Nature*, 24 August 2011, pp. 393–4.

4 Charles Faulkner and Karl Reinhard, 'A Retrospective Examination of Paleoparasitology and Its Establishment in the Journal of Parasitology', *Journal of Parasitology*, C/3 (2014), pp. 253–9; H. Okada, C. Kuhn, H. Feillet and J. Bach, 'The "Hygiene Hypothesis" for Autoimmune and Allergic Diseases: An Update', *Clinical and Experimental Immunology*, CLX/1 (2010), pp. 1–9; T. Olszak et al., 'Microbial Exposure During Early Life has Persistent Effects on Natural Killer T Cell Function', *Nature*, CCCXXXVI/6080 (2012), pp. 489–93; Amirhossein Azari Jafari et al., 'Parasite-Based Interventions in

Systemic Lupus Erythematosus (SLE): A Systematic Review', *Autoimmunity Reviews*, XX/10 (2021), doi: 10.1016/j.autrev.2021.102896.

5 Alfred Crosby, *The Columbian Exchange: Biological and Cultural Consequences of 1492* (Westport, CT, 1972); William McNeill, *Plagues and Peoples* (New York, 1976); John McNeill, *Mosquito Empires: Ecology and War in the Greater Caribbean, 1620–1914* (Cambridge, 2010); George Perkins Marsh, *Man and Nature; or, Physical Geography as Modified by Human Action* [1864] (New York, 1867), p. 106.

6 David S. Jones, 'Virgin Soils Revisited', *William and Mary Quarterly*, LX/4 (2003), pp. 703–42.

7 W.H.S. Jones, *Malaria: A Neglected Factor in the History of Greece and Rome* (London, 1907), p. 96; Burke Cunha, 'The Death of Alexander the Great: Malaria or Typhoid Fever?', *Infectious Disease Clinics of North America*, XVIII/1 (2004), pp. 53–63; James L. A. Webb Jr, *Humanity's Burden: A Global History of Malaria* (Cambridge, 2009), pp. 18–91; Priscila T. Rodrigues et al., 'Human Migration and the Spread of Malaria Parasites to the New World', *Scientific Reports*, VIII/1 (2018), p. 1993, doi: 10.1038/s41598-018-19554-0; Stephen Berger and Stephen Edberg, 'Infectious Diseases in Persons of Leadership', *Reviews of Infectious Diseases*, VI/6 (1984), pp. 802–13; Megan Michel et al., 'Ancient Plasmodium Genomes Shed Light on the History of Human Malaria', *Nature*, DCXXXI (2024), pp. 125–33; Hany Elsheikha, 'Five Deadly Parasites that Have Crossed the Globe', *The Conversation*, https://theconversation.com, 23 May 2015; Leigh Shaw-Taylor, 'An Introduction to the History of Infectious Diseases, Epidemics and the Early Phases of the Long-Run Decline in Mortality', *Economic History Review*, LXXIII/3 (2020), doi: 10.1111/ehr.13019; John Robb et al., 'The Great Health Problem of the Middle Ages? Estimating the Burden of Disease in Medieval England', *International Journal of Paleopathology*, XXXIV (2021), pp. 101–12.

8 Woodrow Borah, 'The Historical Demography of Aboriginal and Colonial America: An Attempt at Perspective', in *The Native Population of the Americas in 1492*, ed. William Denevan (Madison, WI, 1992), jacket cover, p. 30; Crosby, *The Columbian Exchange*; Charles Mann, '1491', *The Atlantic*, March 2002.

9 Philip D. Curtin, 'Disease Exchange across the Tropical Atlantic', *History and Philosophy of the Life Sciences*, XV/3 (1993), pp. 329–56; Peter J. Dowling, '"A Great Deal of Sickness": Introduced Diseases among the Aboriginal People of Colonial Southeast Australia, 1788–1900', PhD thesis, Australian National University, 1997; Brigitte Pakendorf, *Contact in the Prehistory of the Sakha (Yakuts): Linguistic and Genetic Perspectives* (Utrecht, 2007), p. 18; see illustrations at *Visual Representations of the Third Plague Pandemic Photographic Database*, www.repository.cam.ac.uk/handle/1810/281854, accessed 16 May 2025.

10 Toribio Motolinía quoted in Alfred Crosby, 'Conquistador y Pestilencia: The First New World Pandemic and the Fall of the Great Indian Empires', *Hispanic American Historical Review*, XLVII/3 (1967), p. 333.

11 Alan Lymbery et al., 'Co-Invaders: The Effects of Alien Parasites on Native Hosts', *International Journal of Parasitology: Parasites and Wildlife*, III/2 (2014), pp. 171–7; A. Ramenofsky, 'Diseases of the Americas, 1492–1700',

in *The Cambridge World History of Human Disease*, ed. K. Kiple (Cambridge, 1993), pp. 317–28; Jones, 'Virgin Soils Revisited', pp. 703–42; Brenda J. Baker and George J. Armelagos, 'The Origin and Antiquity of Syphilis: Paleopathological Diagnosis and Interpretation', *Current Anthropology*, XXIX/5 (1988), pp. 703–38.

12 McNeill, *Mosquito Empires*. See also Curtin, 'Disease Exchange across the Tropical Atlantic', who explains that 'African diseases . . . set up conditions that made residence in the tropical Americas dangerous for newly arrived people from Europe.'

13 Quote from footnote 8 in Gordon Harrison, *Mosquitoes, Malaria, and Man: A History of the Hostilities since 1880* (New York, 1968), p. 266; Philip Curtin, *Disease and Empire: The Health of European Troops in the Conquest of Africa* (Cambridge, 1998), p. 21; John McNeill, 'How the Lowly Mosquito Helped America Win Independence', *Smithsonian Magazine*, 15 June 2016.

14 Joshua Lederberg, 'Viruses and Humankind: Intracellular Symbiosis and Evolutionary Competition', in *Emerging Viruses*, ed. Stephen S. Morse (New York, 1996), pp. 3–9; Crosby, *The Columbian Exchange*, p. 39; experts quoted in Paul Sutter, '"The First Mountain to Be Removed": Yellow Fever Control and the Construction of the Panama Canal', *Environmental History*, XXI/2 (2016), p. 255.

15 Webb, *Humanity's Burden*, pp. 155–6; Randall Packard, *The Making of a Tropical Disease: A Short History of Malaria* (Baltimore, MD, 2007), pp. 241–2; Denise Doolan, Carlota Dobaño and Kevin Baird, 'Acquired Immunity to Malaria', *Clinical Microbiology Reviews*, XXII/1 (2009), pp. 13–36; Crosby, *The Columbian Exchange*, p. 47.

16 McNeill, *Plagues and Peoples*, p. 52.

17 Bruce Aylward and Rudolf Tangermann, 'The Global Polio Eradication Initiative: Lessons Learned and Prospects for Success', *Vaccine*, 29 (2011), pp. D80–85; Nidia De Jesus, 'Epidemics to Eradication: The Modern History of Poliomyelitis', *Virology Journal*, IV/10 (2007), doi: 10.1186/1743-422X-4-70; Konstantin Chumakov, 'Can Existing Live Vaccines Prevent COVID-19?', *Science*, CCCLXVIII/6496 (2020), pp. 1187–8, doi: 10.1126/science.abc4262. The claim that yellow fever is of African origin is made in Erin Staples and Thomas Monath, 'Yellow Fever: 100 Years of Discovery', *Journal of the American Medical Association*, CCC/8 (2008), pp. 960–62.

18 Frédéric Piel et al., 'Global Distribution of Sickle Cell Gene and Geographical Confirmation of the Malaria Hypothesis', *Nature Communications*, 1/104 (2010), p. 104; Christopher Moxon, 'Malaria: Modification of the Red Blood Cell and Consequences in the Human Host', *British Journal of Haematology*, CLIV/6 (2011), doi: 10.1111/j.1365-2141.2011.08755.x; Pardis Sabeti, 'Natural Selection: Uncovering Mechanisms of Evolutionary Adaptation to Infectious Disease', *Nature Education*, 1/1 (2008), p. 13; Piel et al., 'Global Distribution of the Sickle Cell Gene'; Bridget Penman et al., 'The Emergence and Maintenance of Sickle Cell Hotspots in the Mediterranean', *Infection, Genetics and Evolution*, XII/7 (2012), pp. 1543–50, doi: 10.1016/j.meegid.2012.06.001; A. Allison, 'Protection Afforded by Sickle-Cell Trait against Subtertian Malarial Infection', *British Medical Journal*, 1/4857 (1954), pp. 290–94.

19 M. Siniscalco et al., 'Favism and Thalassaemia in Sardinia and Their
   Relationship to Malaria', *Nature*, CXC/4782 (1961), pp. 1179–80; Timothy
   Winegard, *The Mosquito: A Human History of Our Deadliest Predator* (New
   York, 2019); Mart Stewart, *'What Nature Suffers to Groe': Life, Labor, and
   Landscape on the Georgia Coast, 1680–1920* (Athens, GA, 1996), p. 64; Katherine
   Johnston, *The Nature of Slavery: Environment and Plantation Labor in the Anglo-
   Atlantic World* (Oxford, 2022), p. 217; Linda Nash, *Inescapable Ecologies:
   A History of Environment, Disease, and Knowledge* (Berkeley, CA, 2006), p. 99.
20 *Annual Reports of the War Department for the Fiscal Year Ended June 30, 1898*,
   Report of the Major-General Commanding the Army, vol. III (Washington,
   DC, 1899), p. 515; Marvin Fletcher, 'The Black Volunteers in the Spanish-
   American War', *Military Affairs*, XXXVIII/2 (1974), p. 51; Roger Cunningham,
   '"A Lot of Fine, Sturdy Black Warriors": Texas's African American "Immunes"
   in the Spanish-American War', in *Brothers to the Buffalo Soldiers: Perspectives
   on the African American Militia and Volunteers, 1865–1917*, ed. Bruce Glasrud
   (New York, 2011), pp. 186–208.
21 Johann Wolfgang von Goethe quoted in *Faust*: 'Gewöhnlich glaubt der
   Mensch, wenn er nur Worte hört,/ Es müsse sich dabei doch auch was
   denken lassen.' (Literally: Usually, when a person hears only words, they
   believe that there must be something to think about.)
22 Georges Snounou and Nicolas White, 'The Co-Existence of *Plasmodium*:
   Sidelights from Falciparum and Vivax Malaria in Thailand', *Trends in
   Parasitology*, XX/7 (2004), pp. 333–9; Jetsumon Sattabongkot et al.,
   'Prevalence of Asymptomatic *Plasmodium* Infections with Sub-Microscopic
   Parasite Densities in the Northwestern Border of Thailand: A Potential
   Threat to Malaria Elimination', *Malaria Journal*, XVII/329 (2018), p. 329;
   P. W. Gething et al., 'Modelling the Global Constraints of Temperature on
   Transmission of *Plasmodium falciparum* and *Plasmodium vivax*', *Parasites and
   Vectors*, IV/92 (2011), pmid:21615906; Omar Cornejo and Ananias Escalante,
   'The Origin and Age of *Plasmodium vivax*', *Trends in Parasitology*, XXII/12
   (2007), pp. 558–63; Dorothy Loy et al., 'Out of Africa: Origins and Evolution
   of the Human Malaria Parasites *Plasmodium falciparum* and *Plasmodium
   vivax*', *International Journal for Parasitology*, XLVII/2–3 (2017), pp. 87–97.
23 Margaret Humphreys, *Malaria: Poverty, Race, and Public Health in the United
   States* (Baltimore, MD, 2001); Ivo Mueller et al., 'Natural Acquisition of
   Immunity to *Plasmodium vivax*: Epidemiological Observations and Potential
   Targets', in *Advances in Parasitology: The Epidemiology of Plasmodium Vivax:
   History, Hiatus, and Hubris, Part 2*, ed. S. Hay, Ric Price and Kevin Baird
   (Kidlington, 2013), p. 80; Doolan, Dobaño and Baird, 'Acquired Immunity to
   Malaria', p. 16; Nash, *Inescapable Ecologies*, pp. 22–3.
24 John Frith, 'Syphilis: Its Early History and Treatment Until Penicillin and
   the Debate on Its Origins', *Journal of Military and Veterans' Health*, XX/4
   (2012), pp. 49–58; Robert Kaplan, 'Syphilis, Sex and Psychiatry, 1789–1925:
   Part 2', *Australasian Psychiatry*, XVIII/1 (2010), pp. 22–7; Deborah Hayden,
   Pox: Genius, Madness and the Mysteries of Syphilis (New York, 2003).
25 G. Gartlehner and K. Stepper, 'Julius Wagner-Jauregg: Pyrotherapy,
   Simultanmethode, and "Racial Hygiene"', *James Lind Library Bulletin:
   Commentaries on the History of Treatment Evaluation*, www.jameslindlibrary.org,

accessed 5 May 2015; Cynthia Tsay, 'Julius Wagner-Jauregg and the Legacy of Malarial Therapy for the Treatment of General Paresis of the Insane', *Yale Journal of Biology and Medicine*, LXXXVI/2 (2013), pp. 245–54.

26 K. J. Williams, 'The Introduction of "Chemotherapy" Using Arsphenamine: The First Magic Bullet', *Journal of the Royal Society of Medicine*, CII/8 (2009), pp. 343–8; Jay Geller, *The Other Jewish Question: Identifying the Jew and Making Sense of Modernity* (New York, 2011), p. 100; Julius Wagner-Jauregg, 'Nobel Lecture', 13 December 1927, www.nobelprize.org, accessed 20 June 2015; Eli Chernin, 'The Malariatherapy of Neurosyphilis', *Journal of Parasitology*, LXX/5 (1984), pp. 611–17; Norman Epstein, 'Artificial Fever as a Therapeutic Procedure', *California and Western Medicine*, XLIV/5 (1936), pp. 357–8.

27 Chernin, 'The Malariatherapy of Neurosyphilis', p. 612. For more details about the Tuskegee Syphilis Study, see, for example, Loretta Cormier, *The Ten-Thousand Year Fever: Rethinking Human and Wild-Primate Malaria* (New York, 2011); Harriet A. Washington, *Medical Apartheid: The Dark History of Medical Experimentation on Black Americans from Colonial Times to Present* (New York, 2006), pp. 181–2.

28 Christopher Hamlin, *More Than Hot: A Short History of Fever* (Baltimore, MD, 2014), p. 1.

29 'Heimlich Maneuvers into AIDS Therapy', CNN, http://us.cnn.com, 14 April 2003; H. J. Heimlich et al., 'Malariotherapy for HIV Patients', *Mechanism of Aging and Development*, XC/1–3 (1997), pp. 79–85; Michel Drancourt and Didier Raoult, 'Malaria Therapy for Ebola Virus Infection', *Clinical Infectious Diseases*, LXIV/5 (2017), pp. 696–7; Chernin, 'The Malariatherapy of Neurosyphilis', pp. 611–17; Marc Cohen, 'Turning Up the Heat on COVID-19: Heat as a Therapeutic Intervention', *F1000Research*, IX/292 (July 2020), doi: 10.12688/f1000research.23299.2; René Dubos, *Mirage of Health: Utopias, Progress, and Biological Change* (New York, 1959), p. 88.

30 Sema Nickbakhsh et al., 'Virus-Virus Interactions Impact the Population Dynamics of Influenza and the Common Cold', PNAS, CXVI/52 (2019), doi: 10.1073/pnas.1911083116; Stephen Ell, 'Immunity as a Factor in the Epidemiology of Medieval Plague', *Reviews of Infectious Diseases*, VI/6 (1984), pp. 866–79; Richard Hoffmann, *An Environmental History of Medieval Europe* (Cambridge, 2014), p. 289.

31 Michael Worobey, Guan-Zhu Han and Andrew Rambaut, 'Genesis and Pathogenesis of the 1918 Pandemic H1N1 Influenza Virus', PNAS, CXI/22 (2014), pp. 8107–12.

32 Deepshika Ramanan et al., 'Helminth Infection Promotes Colonization Resistance Via Type 2 Immunity', *Science*, CCCLII/6285 (2016), doi: 10.1126/science.aaf3229; Peter Aaby et al., 'Low Mortality after Mild Measles Infection Compared to Uninfected Children in Rural West Africa', *Vaccine*, XXI/1–2 (2002), pp. 120–26; Che Julius Ngwa and Gabriele Pradel, 'Coming Soon: Probiotics-Based Malaria Vaccines', *Trends in Parasitology*, XXXI/1 (2015), p. 1, doi: 10.1016/j.pt.2014.11.006; Bahtiyar Yilmaz et al., 'Gut Microbiota Elicits a Protective Immune Response against Malaria Transmission', *Cell*, CLIX/6 (2014), pp. 1277–89, doi: 10.1016/j.cell.2014.10.053; Cláudio Daniel-Ribeiro and Graziela Zanini, 'Autoimmunity and Malaria: What Are They Doing Together?', *Acta Tropica*, LXXVI/3 (2000), pp. 205–21.

33 Konstantin Chumakov et al., 'Can Existing Live Vaccines Prevent COVID-19?', *Science*, CCCLXVIII/6496 (2020), pp. 1187–2, doi: 10.1126/science.abc4262; Konstantin Chumakov et al., 'Old Vaccines for New Infections: Exploiting Innate Immunity to Control COVID-19 and Prevent Future Pandemics', *PNAS*, CXVIII/21 (2021), doi: 10.1073/pnas.2101718118.

34 Pierre Joseph van Beneden, *Animal Parasites and Messmates*, 2nd edn (London, 1876), pp. 2, 85.

35 J. Lilue et al., 'Reciprocal Virulence and Resistance Polymorphism in the Relationship Between Toxoplasma Gondii and the House Mouse', *eLife*, II (2013), doi: 10.7554/eLife.01298; Patrick House, Ajai Vyas and Robert Sapolsky, 'Predator Cat Odors Activate Sexual Arousal Pathways in Brains of *Toxoplasma gondii* Infected Rats', *PLOS ONE*, VI/8 (2011), p. e23277, doi:10.1371/journal.pone.0023277.

36 Anand Vasudevan, 'Parasitic Manipulation of Male Sexual Advertisement in *Toxoplasma gondii–Rattus norvegicus* Association', PhD Thesis, Nanyang Technological University, 2016; Øyvind Øverli and Ida Johansen, 'Kindness to the Final Host and Vice Versa: A Trend for Parasites Providing Easy Prey?', *Frontiers in Ecology and Evolution*, VII (2019), doi: 10.3389/fevo.2019.00050.

37 Ailie Robinson et al., '*Plasmodium*-Associated Changes in Human Odor Attract Mosquitoes', *PNAS*, CXV/18 (2018), pp. e4209–18, doi: 10.1073/pnas.1721610115.

38 Jeffrey Jones et al., '*Toxoplasma gondii* Infection in the United States, 2011–2014', *American Journal of Tropical Medicine and Hygiene*, XCVIII/2 (2018), pp. 551–7; Emily Severance et al., '*Toxoplasma gondii*: A Gastrointestinal Pathogen Associated with Human Brain Diseases', *International Review of Neurobiology*, CXXXI (2016), doi: 10.1016/bs.irn.2016.08.008; João Furtado et al., 'Toxoplasmosis: A Global Threat', *Journal of Global Infectious Diseases*, III/3 (2011), pp. 281–4, doi: 10.4103/0974-777X.83536.

39 Alessandra Nicoletti et al., '*Toxoplasma gondii* and Multiple Sclerosis: A Population-Based Case-Control Study', *Scientific Reports*, X/18855 (2020), doi: 10.1038/s41598-020-75830-y; Ann-Kathrin Stock et al., 'Latent *Toxoplasma gondii* Infection Leads to Improved Action Control', *Brain, Behavior, and Immunity*, XXXVII (2014), pp. 103–8; Thomas Cook et al., '"Latent" Infection with *Toxoplasma gondii*: Association With Trait Aggression and Impulsivity in Healthy Adults', *Journal of Psychiatric Research*, LX (2015), pp. 87–94; Stephanie Johnson et al., 'Risky Business: Linking *Toxoplasma gondii* Infection and Entrepreneurship Behaviours across Individuals and Countries', *Proceedings of the Royal Society B*, 285 (2018), doi: 10.1098/rspb.2018.0822; Jaroslav Flegr, 'Influence of Latent *Toxoplasma* Infection on Human Personality, Physiology and Morphology: Pros and Cons of the *Toxoplasma*-Human Model in Studying the Manipulation Hypothesis', *Journal of Experimental Biology*, CCXVI/1 (2013), pp. 127–33, doi: 10.1242/jeb.073635; Clémence Poirotte et al., 'Morbid Attraction to Leopard Urine in Toxoplasma-Infected Chimpanzees', *Current Biology*, XXVI/3 (2016), p. R98, doi: 10.1016/j.cub.2015.12.020; Sarka Kanková et al., 'Women Infected with Parasite *Toxoplasma* Have More Sons', *Naturwissenschaften*, XCIV/2 (2007), pp. 122–7, doi: 10.1007/s00114-006-0166-2; Vinita Ling et al.,

'*Toxoplasma gondii* Seropositivity and Suicide Rates in Women', *Journal of Nervous and Mental Disease*, CXCIX/7 (2011), doi: 10.1097/ NMD.0b013e318221416e.

40 Candace Williams et al., 'Regulation of Endocrine Systems by the Microbiome: Perspectives from Comparative Animal Models', *General and Comparative Endocrinology*, CCXCII (2020), doi: 10.1016/j.ygcen.2020.113437.

41 Michael Pollan, *The Botany of Desire: A Plant's-Eye View of the World* (New York, 2001); N. Stanczyk et al., 'Species-Specific Alterations in Anopheles Mosquito Olfactory Responses Caused by Plasmodium Infection', *Scientific Reports*, IX/1 (2019), p. 3396, doi: 10.1038/S41598-019-40074-Y; Cédric Sueur and Michael Huffman, 'Co-Cultures: Exploring Interspecies Culture Among Humans and Other Animals', *Trends in Ecology and Evolution*, XXXIX/9 (2024), pp. 821–9.

42 Arthur Talman et al., 'Uptake of *Plasmodium falciparum* Gametocytes during Mosquito Bloodmeal by Direct and Membrane Feeding', *Frontiers of Microbiology*, XI (2020), doi: 10.3389/fmicb.2020.00246.

43 Reinhard Hoeppli and I-Hung Ch'iang, 'Selections from Old Chinese Medical Literature on Various Subjects of Helminthological Interest', *Chinese Medical Journal*, LVII/4 (1940), pp. 374, 376; Julie Hayden Grissom, 'Parasitic Worms in Early Modern Science and Medicine, 1650–1810', PhD Dissertation, University of Oklahoma, 2014; Reinhard Hoeppli, *Parasites and Parasitic Infections in Early Medicine and Science* (Singapore, 1959), pp. 62, 164–5; Forney Zacharias quoted in I. S. Whitaker et al., 'Larval Therapy from Antiquity to the Present Day: Mechanisms of Action, Clinical Applications and Future Potential', *Postgraduate Medical Journal*, LXXXIII/980 (2007), pp. 409–13; Gerry Greenstone, 'The History of Bloodletting', BC *Medical Journal*, LII/1 (2010), pp. 12–14; Stefan Schatzki, 'Washington in His Last Illness Attended by Drs. Craik and Brown', *American Journal of Roentgenology*, CLXVIII/2 (1997), p. 332.

44 Hoeppli, *Parasites and Parasitic Infections*, pp. 164–5; van Beneden, *Animal Parasites and Messmates*, pp. 89–90; Sarah Schenck and Steve Lawrence, *The Invisible Extinction* (2022, 80 mins), rocofilms.com/films, quote at 58 mins; Martin Blaser, *Missing Microbes: How the Overuse of Antibiotics is Fueling Our Modern Plagues* (New York, 2014).

45 Hans Zinsser, *Rats, Lice and History* (London, 1935), p. 187; J. Maunder, 'The Appreciation of Lice', *Proceedings of the Royal Institution of Great Britain*, LV (1983), pp. 1–31; Joseph A. Jackson et al., 'Immunomodulatory Parasites and Tool-Like Receptor-Mediated Tumour Necrosis Factor Alpha Responsiveness in Wild Mammals', BMC *Biology*, VII/16 (2009), doi: 10.1186/1741-7007-7-16.

46 'Structure, Function, and Diversity of the Human Microbiome: Human Microbiome Project Consortium', *Nature*, CDLXXXVI (2012), pp. 207–14.

47 Edwin Robinson, 'Notes on the Life History of *Leucochloridium Fuscostriatum N. Sp. Provis.* (Trematoda: Brachylaemidae)', *Journal of Parasitology*, XXXIII/6 (1947), pp. 467–75.

48 Robert Poulin, *Evolutionary Ecology of Parasites* (London, 1998), pp. 80–81; Joe Pierre, 'Fear the Walking Dead: Can Brain Parasites Make Us Zombies?', *Psychology Today*, 28 September 2015; F.E.G. Cox,

'History of Human Parasitology', *Clinical Microbiology Reviews*, XV/4 (2002), pp. 595–612, doi: 10.1128/CMR.15.4.595-612.2002; Kevin Lafferty, 'Can the Common Brain Parasite, *Toxoplasma gondii*, Influence Human Culture?', *Proceedings of the Royal Society B*, CCLVVIII/1602 (2006), doi: 10.1098/rspb.2006.3641.

49 Walter Scheidel, *The Great Leveler: Violence and the History of Inequality from the Stone Age to the Twenty-First Century* (Princeton, NJ, 2017), p. 335; McNeill, *Plagues and Peoples*, p. 219.

50 Sharon DeWitte, 'The Anthropology of Plague: Insights from Bioarcheological Analyses of Epidemic Cemeteries', *Medieval Globe*, I/1 (2014), pp. 97–123; Kathryn Olivarius, 'Necropolis: Yellow Fever, Immunity, and Capitalism in the Deep South, 1800–1860', DPhil Thesis, Oxford University, 2016, p. 13; Jonathan Kennedy, *Pathogenesis: A History of the World in Eight Plagues* (New York, 2023).

4 Dreams of Eradication

1 James S. Brady, 'Remarks by President Trump', Briefing Statements, www.whitehouse.gov, 19 March 2020; Lev Facher, 'Fact-Checking Trump's Claims about Hydroxychloroquine, the Antimalarial Drug He's Touting as a Coronavirus Treatment', www.statnews.com, 6 April 2020; Heidi Ledford, 'Chloroquine Hype Is Derailing the Search for Coronavirus Treatments', *Nature*, DLXXX/7805 (2020), p. 573.

2 Donald J. Trump on Twitter @realDonaldTrump, 21 March 2020; Dr Dena Grayson @DrDenaGrayson replying to @realDonaldTrump, 21 March 2020 on Twitter; Mayo Clinic, 'Hydroxychloroquine (Oral Route)', www.mayoclinic.org, accessed 21 August 2020; Mayla Borba, Fernando Val and Vanderson Sampaio, 'Effect of High vs Low Doses of Chloroquine Diphosphate as Adjunctive Therapy for Patients Hospitalized with Severe Acute Respiratory Syndrome Coronavirus 2 (SARS-CoV-2) Infection', *JAMA*, III/4 (2020), p. e208857; James S. Brady, 'Remarks by President Trump', Briefing Statements, www.whitehouse.gov, 23 April 2020; Lindzi Wessel, '"It's a nightmare." How Brazilian Scientists Became Ensnared in Chloroquine Politics', *Science*, 22 June 2020.

3 Martin J. Vincent et al., 'Chloroquine Is a Potent Inhibitor of SARS Coronavirus Infection and Spread', *Virology Journal*, II/69 (2005), doi: 10.1186/1743-422X-2-69; Guanguan Li et al., 'Enantiomers of Chloroquine and Hydroxychloroquine Exhibit Different Activities against SARS-CoV-2 *In Vitro*, Evidencing S-Hydroxychloroquine as a Potentially Superior Drug for COVID-19', *bioRxiv* [not peer-reviewed]; the controversial retracted study was Mandeep Mehra et al., 'Hydroxychloroquine or Chloroquine With or Without a Macrolide for Treatment of COVID-19: A Multinational Registry Analysis', *The Lancet* (May 2020), doi: 10.1016/S0140-6736(20)31180-6.

4 G. R. Coatney, 'Pitfalls in a Discovery: The Chronicle of Chloroquine', *American Journal of Tropical Medicne and Hygiene*, XII (1963), pp. 121–8.

5 Kim Walker and Mark Nesbitt, *Just the Tonic: A Natural History of Tonic Water* (Kew, 2020); William Harty, 'On the Use of Quinine in the Ague of Dublin and Its Power in Accelerating Mercurial Action', *Edinburgh*

Medical and Surgical Journal, XXXII/101 (1829), p. 338; James L. A. Webb Jr,
Humanity's Burden: A Global History of Malaria (Cambridge, 2009),
pp. 175–7; Elisabeth Hsu, 'Reflections on the "Discovery" of the Antimalarial
Qinghao', *British Journal of Clinical Pharmacology*, LXI/6 (2006), pp. 666–70.

6 W. C. MaClean, 'Malarial Fevers', in *A System of Medicine*, vol. I, ed. John
Russell Reynolds (Philadelphia, PA, 1868), p. 78; 'Potential Dangers of
Homemade Tonic Water', 4 August 2014 ('Coronavirus Update March 28,
2020'), www.alcademics.com; Jacqui Wise, 'Long Term Quinine for Muscle
Cramps May Increase Death Risk', *British Medical Journal*, CCCLVII (2017),
doi: 10.1136/bmj.j2236; Matt Spinozzi, 'Cinchona Bark', *Distiller*, https://
distilling.com, 31 August 2020.

7 Matthew Crawford, *The Andean Wonder Drug: Cinchona Bark and Imperial
Science in the Spanish Atlantic, 1630–1800* (Pittsburgh, PA, 2016); Samuel
Stanley, 'Antiparasitic Agents', in *Infectious Diseases*, ed. Jonathan Cohen,
Steven Opal and William Powderly (London, 2010), ch. 150; Debra Woods
and Tracey Williams, 'The Challenges of Developing Novel Antiparasitic
Drugs', *Invertebrate Neuroscience*, VII/4 (2007), pp. 245–50.

8 Mark Honigsbaum, *The Fever Trail: In Search of the Cure for Malaria* (New
York, 2001), p. 23; Zhang Jianfang, *A Detailed Chronological Record of Project
523 and the Discovery and Development of Qinghaosu (Artemisinin)*, trans.
Muoi and Keith Arnold (Houston, TX, 2013); 'Artemisinin and Traditional
Chinese Medicine (TCM)', *Target Health LLC*, www.targethealth.com,
2 December 2019.

9 Merlin Wilcox and Gerard Bodeker, 'Traditional Herbal Medicines
for Malaria', *British Medical Journal*, CCCXXIX/7475 (2004), pp. 1156–9,
doi: 10.1136/bmj.329.7475.1156; I. Cock, M. Selesho and S. van Vuuren,
'A Review of the Traditional Use of Southern African Medicinal Plants
for the Treatment of Malaria', *Journal of Ethnopharmacology*, CCXLV (2019),
112176, doi: 10.1016/j.jep.2019.112176; N. Nundkumar and J. A. Ojewole,
'Studies on the Antiplasmodial Properties of Some South African Medicinal
Plants Used as Antimalarial Remedies in Zulu Folk Medicine', *Methods and
Findings in Experimental and Clinical Pharmacology*, XXIV/7 (2002),
pp. 397–401; Hermione Manekeng et al., 'Evaluation of Acute and Subacute
Toxicities of *Psidium guajava* Methanolic Bark Extract: A Botanical with
In Vitro Antiproliferative Potential', *Evidence-Based Complementary and
Alternative Medicine*, 4 (2019), doi: 10.1155/2019/8306986.

10 Interview with Antonio Melis (born 1928) by the fifth-grade class of Laconi,
Sardinia, 2000 (Pierluigi Cocco papers, Cagliari, 2000); Giuseppe Tagarelli,
Antonio Tagarelli and Anna Piro, 'Folk Medicine Used to Heal Malaria in
Calabria (Southern Italy)', *Journal of Ethnobiology and Ethnomedicine*, VI/27
(2010), doi: 10.1186/1746-4269-6-27.

11 Franco Bonelli, 'La malaria nella storia demografica ed economica d'Italia:
Primi lineamenti di una ricerca', *Studi storici*, VII/4 (1966), p. 663; Luigi
Torelli, *Carta della malaria dell'Italia* (Florence, 1882), p. 29; Denis Mack
Smith, *Italy: A Modern History* (Ann Arbor, MI, 1959), p. 494.

12 Tommaso La Mantia, 'Storia dell'Eucalitticoltura in Sicilia', *Il naturalista
siciliano*, IV/37 (2013), pp. 587–628; Arrigo Serpieri, *La Bonifica nella storia
e nella dottrina* (Bologna, 1957), p. 104; Marcus Hall, *Earth Repair:*

A Transatlantic History of Environmental Restoration (Charlottesville, VA, 2005), pp. 161–91.

13 Frank Snowden, *The Conquest of Malaria: Italy, 1900–1962* (New Haven, CT, 2006), pp. 53–86; Giuseppe Tropeano, *La malaria nel mezzogiorno d'Italia* (Naples, 1908), pp. 333, 366, 368; Frank Snowden, 'The Use and Misuse of History: Lessons from Sardinia', in *The Global Challenge of Malaria: Past Lessons and Future Prospects,* ed. Frank Snowden and Richard Bucala (Singapore, 2014), p. 77; 'La lotta contro la Malaria' (1909) quoted in Sergio Sabbatani and Antonio Sandri, 'La profilassi chininica agli inizi del XX secolo in Italia e nell'area a nord di Bologna teatro di una epidemia malarica', *Le infezioni in medicina,* III (2000), pp. 176–90.

14 André Felipe Câdido da Silva and Jaime Larry Benchimol, 'Malaria and Quinine Resistance: A Medical and Scientific Issue between Brazil and Germany (1907–19)', *Medical History,* LVIII/1 (2014), pp. 1–26.

15 Vincenzo Medde, 'Uomini, terre e malaria in Sardegna dal Settecento al Fascismo', *IcoNur: Vediamo ciò che sappiamo,* www.iconur.it, accessed 26 August 2021; Teodor Kaufman and Edmundo Rúveda, 'The Quest for Quinine: Those Who Won the Battles and Those Who Won the War', *Angewandte Chemie International Edition,* XLIV/6 (2005), pp. 854–85; L. J. Bruce-Chwatt, 'Clinical Trials of Antimalarial Drugs', *Trans. Royal Society of Tropical Medicine and Hygiene,* LXI/3 (1967), p. 412; W. Schulemann, 'Synthetic Anti-Malarial Preparations', *Proceedings of the Royal Society of Medicine,* XXV/6 (1932), p. 902; Giancarlo Majori and Federica Napolitani, eds, *Il laboratorio di malariologia* (Rome, 2010), p. 34.

16 Robert Kaplan, 'Syphilis, Sex and Psychiatry, 1789–1925: Part 2', *Australasian Psychiatry,* XVIII/1 (2010), p. 23.

17 Reginald Manwell, 'Malaria, Birds, and War', *American Scientist,* XXXVII/1 (1949), pp. 60–68; 'Ethics Governing the Service of Prisoners as Subjects in Medical Experiments', Report of a Committee Appointed by Governor Dwight H. Green of Illinois, *Journal of the American Medical Association,* CXXXVI/7 (1948), p. 458.

18 H.B.F. Dixon, 'A Report on Six Hundred Cases of Malaria Treated with Plasmoquine and Quinine', *Journal of the Royal Army Medical Corps,* LX/6 (1933), pp. 431–9; Quote from 'Book Notices: *Malaria in Europe: An Ecological Study,* by L. W. Hackett (1937)', *Journal of the American Medical Association,* CXI/8 (August 1938), p. 745; H. Weniger, 'Toxicity and Side Effects of Primaquine and other 8-Aminoquinolines', World Health Organization WHO/MAL/79.905 (1979), https://apps.who.intf, accessed 26 August 2021; 'Plasmoquine Capsule', www.tabletwise.net, accessed 8 June 2023.

19 Carlo Levi, *Christ Stopped at Eboli* [1945], trans. Frances Frenaye (New York, 1947), p. 242.

20 Daniel J. Wallace, 'The Use of Quinacrine (Atabrine) in Rheumatic Diseases: A Reexamination', *Seminars in Arthritis and Rheumatism,* XVIII/4 (1989), p. 218; Peter Weina, 'From Atabrine in World War II to Mefloquine in Somalia: The Role of Education in Preventive Medicine', *Military Medicine,* CLXIII/9 (1998), pp. 635–9.

21 *Adventure in Sardinia* (1950, 20 mins) dir. Peter Baylis and produced by British Pathé, quoted in Marcus Hall, 'World War II and the Axis of Disease',

in *War and the Environment: Military Destruction in the Modern Age,* ed. Charles Closmann (College Station, TX, 2009), p. 118; Marcus Hall, 'Today Sardinia, Tomorrow the World: Malaria, the Rockefeller Foundation, and Mosquito Eradication', *BardPolitik: The Bard Journal of Global Affairs,* 5 (Fall 2004), at www.issuelab.org.

22   For a global survey of malaria control, see James L. A. Webb Jr, *Humanity's Burden: A Global History of Malaria* (Cambridge, 2009); on some of the pre-DDT methods for controlling malaria in Sardinia, see E. Tognotti, 'Malaria in Sardinia', *International Journal of Anthropology,* XIII/3–4 (1998), pp. 237–42.

23   M. Fua, 'Sopra l'azione di alcuni insetticidi', *Giornale di Agricoltura Pratica,* V (1894), p. 149; Carlo Tonini, 'Manuale di Chimica Tecnologica (Parte Seconda)', *Memorie dell'Accademia d'Agricoltura Commercio ed Arti di Verona,* 227 (1851), p. 524; Wei Bao et al., 'Association Between Exposure to Pyrethroid Insecticides and Risk of All-Cause and Cause-Specific Mortality in the General U.S. Adult Population', *JAMA Intern Med.,* CLXXX/3 (2020), pp. 367–74; 'Summary 1, Neocid', 11 May 1943, quoted in Edmund Russell III, 'The Strange Career of DDT: Experts, Federal Capacity, and Environmentalism of World War II', *Technology and Culture,* XL/4 (1999), p. 780.

24   Sarah Moore and Mustapha Debboun, 'History of Insect Repellents', in *Insect Repellents: Principles, Methods, and Uses,* ed. M. Debboun, S. P. Frances and D. Strickman (Boca Raton, LA, 2007), pp. 3–30.

25   Gordon Patterson, 'Looking Backward, Looking Forward: The Long, Torturous Struggle with Mosquitoes', *Insects,* VII/56 (2016), pp. 1–14; Hall, 'World War II and the Axis of Disease', pp. 118–19; James L. A. Webb Jr, *The Long Struggle Against Malaria in Tropical Africa* (Cambridge, 2014), p. 77, footnote 6; Jennifer Saxe, Teresa Bowers and Kim Reid, 'Arsenic', in *Environmental Forensics,* ed. Robert Morrison and Brian Murphy (London, 2006), pp. 279–92; 'Story of Mrs Luce's Poisoning in Rome Villa Amazes Italians', *New York Times,* 18 July 1956, p. 29.

26   Marston Bates quoted in Marcus Hall and Dan Tamir, 'Killing Mosquitoes: Think Before You Swat', in *Mosquitopia: The Place of Pests in a Healthy World,* ed. Marcus Hall and Dan Tamïr (London, 2021), p. 5; *The Sardinian Project* (1949, 33 mins), produced by Arthur Elton and directed by Jack Chambers, Shell Film Unit, UK.

27   'DDT: Technical Fact Sheet', National Pesticide Information Center (2000), http://npic.orst.edu/factsheets/archive/ddttech.pdf, accessed 1 September 2021; Pierluigi Cocco and Domenica Fadda, 'Cancer Mortality Among Men Occupationally Exposed to Dichlorodiphenyltrichloroethane', *Cancer Research,* LXV/20 (2005), pp. 9588–94, doi: 10.1158/0008-5472. CAN-05-1487. See for example, 'Tumori, ISTAT: 'In Sardegna e Campania il tasso di mortalità più alto d'Italia', *L'Unione Sarda,* www.unionesarda.it, 14 June 2018.

28   Marcus Hall, 'The Rockefeller Foundation in Sardinia: Pesticide Politics in the Struggle Against Malaria', Working Paper at the Carnegie Council on Ethics and International Affairs (2004), New York, 28 March 2005, at www.carnegiecouncil.org; Peter J. Brown, 'Cultural Adaptations to Endemic Malaria and the Socioeconomic Effects of Malaria Eradication in Sardinia', PhD thesis, State University of New York at Stony Brook, 1979.

29 Olivia Orta et al., 'Correlates of Organochlorine Pesticide Plasma
    Concentrations Among Reproductive-Aged Black Women', *Environmental
    Research*, CLXXXIV (2020), doi: 10.1016/j.envres.2020.109352.

30 Environmental Protection Agency, 'Controlling Adult Mosquitoes',
    www.epa.gov, accessed 2 September 2021; Giovanni Benelli, Claire
    Jeffries and Thomas Walker, 'Biological Control of Mosquito Vectors:
    Past, Present, and Future', *Insects*, LXXIV/4 (2016), pp. 52; Alina Kulman
    and Dan Tamir, 'A Man and His Minnows: The Introduction of *Gambusia
    affinis* to Mandatory Palestine', *Frontiers in Conservation Science*, III (2022),
    doi: 10.3389/fcosc.2022.649955; Després Laurence, Lagneau Christophe
    and Frutos Roger, 'Using the Bio-Insecticide *Bacillus Thuringiensis Israelensis*
    in Mosquito Control', in *Pesticides in the Modern World: Pests Control and
    Pesticides Exposure and Toxicity Assessment*, ed. Margarita Stoytcheva (Rijeka,
    2011), pp. 93–126; Carsten Brühl et al., 'Environmental and Socioeconomic
    Effects of Mosquito Control in Europe Using the Biocide *Bacillus
    Thuringiensis* Subsp. Israelensis (Bti)', *Science of the Total Environment*,
    DCCXXIV (2020), doi: 10.1016/j.scitotenv.2020.137800.

31 Rosalind E. Howes et al., 'G6PD Deficiency Prevalence and Estimates of
    Affected Populations in Malaria Endemic Countries: A Geostatistical
    Model-Based Map', *PLOS Medicine*, IX/11 (2012), doi: 10.1371/journal.
    pmed.1001339; Gabriella De Vita et al., 'Two Point Mutations Are
    Responsible for G6PD Polymorphism in Sardinia', *American Journal of
    Human Genetics*, XLIV/2 (1989), pp. 233–40.

32 'Malaria', *ISsalute: Informarsi Conoscere Scegliere* (updated 2018),
    www.issalute.it, accessed 2 February 2021; Centers for Disease Control
    and Prevention, 'Choosing a Drug to Prevent Malaria' (updated 2018),
    www.cdc.gov, accessed 7 September 2021; Amy Maxmen, 'Scientists Hail
    Historic Malaria Vaccine Approval – But Point to Challenges Ahead', *Nature*,
    8 October 2021; Damien Hirst, *Lullaby Spring*, permanent exhibit, Museum
    Brandhorst, Munich, Germany; Martin Blaser, *Missing Microbes: How the
    Overuse of Antibiotics is Fueling Our Modern Plagues* (Basingstoke, 2015).

33 Centre for Health Population, The Government of the Hong Kong Special
    Administrative Region, 'Hypertension', www.chp.gov.hk, accessed 11 May
    2020; 'Worldwide Trends in Blood Pressure from 1975 to 2015: A Pooled
    Analysis of 1479 Population-Based Measurement Studies with 19.1 Million
    Participants', *The Lancet*, CCCLXXXIX/10064 (2017), pp. 37–55, doi: 10.1016/
    S0140-6736(16)31919-5.

34 Leandro Silva et al., 'New Concepts in Malaria Pathogenesis: The Role of
    the Renin-Angiotensin System', *Frontier in Cellular and Infection
    Microbiology*, V/103 (2016), doi: 10.3389/fcimb.2015.00103; J. Hunter Young,
    'Evolution of Blood Pressure Regulation in Humans', *Current Hypertension
    Reports*, IX/1 (2007), pp. 13–18; Hans Zinsser, *Rats, Lice and History*
    (London, 1935), p. 187.

35 D. Lincicome, R. Rossan and W. Jones, 'Growth of Rats Infected with
    Trypanosoma lewisi', *Experimental Parasitology*, XIV/1 (1963), pp. 54–65;
    Catherine Burns, Brett Goodwin and Richard Ostfeld, 'A Prescription
    for Longer Life? Bot Fly Parasitism of the White-Footed Mouse', *Ecology*,
    LXXXVI/3 (2005), p. 754.

36 Boon-Peng Hoh, Thuhairah Abdul Rahman and Khalid Yusoff, 'Natural Selection and Local Adaptation of Blood Pressure Regulation and Their Perspectives on Precision Medicine in Hypertension', *Hereditas*, CLVI/1 (2019), doi: 10.1186/s41065-019-0080-1.

37 George Armelagos, 'The Slavery Hypertension Hypothesis – Natural Selection and Scientific Investigation: A Commentary', *Transforming Anthropology*, XIII/2 (2005), pp. 119–24; Julio Gallego-Delgado and Ana Rodriguez, 'Malaria and Hypertension: Another Co-Evolutionary Adaptation?', *Frontiers in Cellular and Infection Microbiology*, IV/121 (2014), doi: 10.3389/fcimb.2014.00121; 'Biotech Dives into Sardinia Gene Pool for Secret of Long Life', *Financial Times*, 18 July 2016; Robert Desowitz, *New Guinea Tapeworms and Jewish Grandmothers: Tales of Parasites and People* (New York, 1981), p. 130.

38 Allan Kalungi et al., 'Less Severe Cases of COVID-19 in Sub-Saharan Africa: Could Co-Infection or a Recent History of *Plasmodium falciparum* Infection Be Protective?', *Frontiers in Immunology*, XVIII/12 (2021), doi: 10.3389/fimmu.2021.565625. See also Silas Acheampong Osei et al., 'Low Incidence of COVID-19 Case Severity and Mortality in Africa: Could Malaria Co-Infection Provide the Missing Link?', *BMC Infectious Diseases*, XXII/1 (2022), doi: 10.1186/s12879-022-07064-4; Stephanie Nolen, 'Trying to Solve a Covid Mystery: Africa's Low Death Rates', *New York Times*, 23 March 2022.

39 Wendy Orent, 'Will the Coronavirus Evolve to Be Less Deadly?' *Smithsonian Magazine*, www.smithsonianmag.com, 16 November 2020; Samuel Alizon and Mircea Sofonea, 'SARS-CoV-2 Virulence Evolution: Avirulence Theory, Immunity and Trade-Offs', *Journal of Evolutionary Biology*, XXXIV/12 (2021), pp. 1–11, doi: 10.1111/jeb.13896.

40 Alizon and Sofonea, 'SARS-CoV-2 Virulence Evolution', p. 4.

41 Ibid., p. 6.

42 René Dubos, *The Mirage of Health: Utopias, Progress, and Biological Change* (New York, 1959), p. 90.

43 Nathan Wolfe, Claire Dunavan and Jared Diamond, 'Origins Of Major Human Infectious Diseases', *Nature*, CDXLVII/7142 (2007), pp. 279–83, doi: 10.1038/nature05775.

44 William Cronon, 'The Uses of Environmental History', *Environmental History Review*, XVII/3 (1993), p. 10.

5 Thinking Like a Parasite

1 Johannes Sommerfeld, 'Plagues and Peoples Revisited', *EMBO Reports*, IV/Suppl 1 (2003), p. s32–4; Ronald Weinstein and Michael Holcomb, 'Reading List: Select Healthcare Transformation Library 2.0', *Telemedicine and e-Health*, XXVII/9 (2021), doi: 10.1089/tmj.2020.0399.

2 William McNeill, *Plagues and Peoples* (New York, 1976), p. 7.

3 Ibid., p. 9; Alfred Crosby, 'Conquistador y Pestilencia: The First New World Pandemic and the Fall of the Great Indian Empires', *Hispanic American Historical Review*, XLVII/3 (1967), p. 322.

4 Paolo Pio and Joachim Frey, 'Pathogenicity, Population Genetics and Dissemination of *Bacillus anthracis*', *Infection, Genetics and Evolution*, LXIV (2018), pp. 115–25, doi: 10.1016/j.meegid.2018.06.024.

5 McNeill, *Plagues and Peoples*, p. 20; Fernando Pedraza, Hanlun Liu, Klementyna A. Gawecka and Jordi Bascompte, 'The Role of Indirect Effects in Coevolution as Mutualism Transitions into Antagonism', *bioRxiv* (2021), doi: 10.1101/2021.11.22.469544.

6 Hans Bremermann and John Pickering, 'A Game-Theoretical Model of Parasite Virulence', *Journal of Theoretical Biology*, C/3 (1983), pp. 411–26. See also Andrew Read and Paul Harvey, 'The Evolution of Virulence', *Nature*, CCCLXII (1993), pp. 500–501.

7 Camille Bonneaud et al., 'Evolution of Both Host Resistance and Tolerance to an Emerging Bacterial Pathogen', *Evolution Letters*, III/5 (2019), doi: 10.1002/evl3.133.

8 Michel Serres, *The Parasite* [1980], trans. Lawrence Schehr (Baltimore, MD, 1982), pp. 15–16.

9 Valeria Salvaterra, '*Para-sitos*: Notes on Incorporation and Hospitality in Jacques Derrida's Unpublished Seminar *Manger l'autre* (1989–1990)', *Philosophy Today*, LXV/3 (2021), pp. 637–53, doi: 10.5840/philtoday2021520411; see also Eben Kirksey, 'Living with Parasites in Palo Verde National Park', *Environmental Humanities*, I/1 (2012), pp. 23–55.

10 McNeill, *Plagues and Peoples*, p. 61; Arturo Casadevall and Liise-anne Pirofski, 'Microbiology: Ditch the Term Pathogen', *Nature*, DXVI (2014), pp. 165–6.

11 Stephanie Johnson et al., 'Risky Business: Linking *Toxoplasma gondii* Infection and Entrepreneurship Behaviours Across Individuals and Countries', *Proceedings of the Royal Society B*, CCLXXXV/1983 (2018), doi: 10.1098/rspb.2018.0822.

12 McNeill, *Plagues and Peoples*, p. 10; Jared Diamond, *Guns, Germs, and Steel: The Fates of Human Societies* (New York, 1997), p. 199.

13 H. Hurd et al., 'Evaluating the Costs of Mosquito Resistance to Malaria Parasites', *Evolution*, LIX/12 (2005), pp. 2560–72; J. Vézilier, A. Nicot, S. Gandon and A. Rivero, 'Plasmodium Infection Decreases Fecundity and Increases Survival of Mosquitoes', *Proceedings of the Royal Society B*, CCLXXIX/1744 (2012), doi: 10.1098/rspb.2012.1394; Josué Martínez-de la Puente, Rafael Gutiérrez-López and Jordi Figuerola, 'Do Avian Malaria Parasites Reduce Vector Longevity?', *Current Opinion in Insect Science*, XXVIII (2018), pp. 113–17.

14 Ailie Robinson et al., 'Plasmodium-Associated Changes in Human Odor Attract Mosquitoes', *Proceedings of the National Academy of Science*, CXV/18 (2018), pp. e4209–18, doi: 10.1073/pnas.1721610115.

15 Yang Zhao et al., 'Enhanced Survival of Plasmodium-Infected Mosquitoes during Starvation', *PLOS One*, VII/7 (2012), p. e40556, doi: 10.1371/journal.pone.0040556; Amélie Vantaux et al., 'Field Evidence for Manipulation of Mosquito Host Selection by the Human Malaria Parasite, *Plasmodium falciparum*', *Peer Community Journal*, I/1430 (2021), p. e13.

16 W. Robert Shaw, Perrin Marcenac and Flaminia Catteruccia, 'Plasmodium Development in Anopheles: A Tale of Shared Resources', *Trends in Parasitology*, XXXVIII/2 (2022), pp. 124–35; James L. A. Webb Jr, 'Early Malarial Infections and the First Epidemiological Transition', in *Human Dispersal and Species Movement: From Prehistory to the Present*, ed. Nicole Boivin, Rémy

Crassard and Michael Petraglia (Cambridge, 2017), pp. 477–93; Clay Huff, 'A Consideration of the Problem of Evolution of Malarial Parasites', *Revista del Instituto de Salubridad y Enfermedades Tropicales*, VI/4 (1945), pp. 253–8.

17 Quote from Gordon Harrison, *Mosquitoes, Malaria, and Man: A History of the Hostilities since 1880* (New York, 1978), p. 266, footnote 8; John McNeill, 'How the Lowly Mosquito Helped America Win Independence', *Smithsonian Magazine*, 15 June 2016; C. G. Huff, 'Studies on the Evolution of Some Disease-Producing Organisms', *Quarterly Review of Biology*, XIII/2 (1938), pp. 196–206; Philip D. Curtin, 'The End of the "White Man's Grave"? Nineteenth-Century Mortality in West Africa', *Journal of Interdisciplinary History*, XXI/1 (1990), pp. 63–88; Ramon Muntaner and Pietro IV d'Aragona, *La Conquista della Sardegna nelle Cronache Catalane*, ed. Giuseppe Meloni (Nuoro, 1999), p. 155.

18 Fletcher Halliday et al., 'On the Hunt for Facilitation in Symbiont Communities', *Trends in Ecology and Evolution*, XXXIX/9 (2024), pp. 793–6; Donna Haraway, *The Companion Species Manifesto: Dogs, People, and Significant Otherness* (Chicago, IL, 2003); Gregg Mitman and Rob Nixon, 'A Dialogue on Form, Knowledge, and Representation', in 'Minding the Gap: Working Across Disciplines in Environmental Studies', RCC *Perspectives*, II (2014), p. 61.

19 André Wilke, Giovanni Benelli and John Beier, 'Beyond Frontiers: on Invasive Alien Mosquito Species in America and Europe', PLOS *Neglected Tropical Diseases*, XIV/1 (2020), doi: 10.1371/journal.pntd.0007864; Ranjan Ramasamy, 'Zoonotic Malaria – Global Overview and Research and Policy Needs', *Frontiers in Public Health*, II/123 (2014), doi: 10.3389/fpubh.2014.00123; Mario Coluzzi, 'The Clay Feet of the Malaria Giant and Its African Roots: Hypotheses and Inferences About Origin, Spread and Control of *Plasmodium falciparum*', *Parassitologia*, XLI/1–2 (1999), pp. 277–83.

20 Marcus Hall and Dan Tamir, 'Killing Mosquitoes: Think before You Swat', in *Mosquitopia: The Place of Pests in a Healthy World*, ed. Marcus Hall and Dan Tamïr (London, 2021), p. 9; Noah Rose et al., 'Climate and Urbanization Drive Mosquito Preference for Humans', *Current Biology*, XXX/18 (2020), doi: 10.1016/j.cub.2020.06.092; Allen Jennings quoted in Paul Sutter, 'Nature's Agents or Agents of Empire? Entomological Workers and Environmental Change during the Construction of the Panama Canal', *ISIS*, XCVIII/4 (2007), pp. 724–54; M. E. Sinka et al., 'A New Malaria Vector in Africa: Predicting the Expansion Range of *Anopheles stephensi* and Identifying the Urban Populations at Risk', PNAS, CXVII/40 (2020), pp. 24900–908, doi: 10.1073/pnas.2003976117. See also Randall Packard, *The Making of a Tropical Disease: A Short History of Malaria* (Baltimore, MD, 2010).

21 Michael Purugganan, 'What Is Domestication?', *Trends in Ecology and Evolution*, 2996 (2022), doi: 10.1016/j.tree.2022.04.006.

22 Michael Pollan, *Botany of Desire: A Plant's-Eye View of the World* (New York, 2001); Haraway, *The Companion Species Manifesto*, p. 53.

23 Urmi Engineer Willoughby, 'Domesticated Mosquitoes: Colonization and Growth of Mosquito Habitats in North America', in *Mosquitopia*, ed. Hall and Tamïr, pp. 66–7.

24 Chris Fonvielle, 'A Plague Most Deadly', *Salt Magazine*, www.saltmagazine nc.com, 30 August 2019.

25 J. Harper and S. Paulson. 'Reproductive Isolation between Florida Strains
    of *Aedes aegypti* and *Aedes albopictus*', *Journal of the American Mosquito
    Control Association*, x/1 (1994), pp. 88–92; Kathryn Hanley et al., 'Fever
    Versus Fever: The Role of Host and Vector Susceptibility and Interspecific
    Competition in Shaping the Current and Future Distributions of the
    Sylvatic Cycles of Dengue Virus and Yellow Fever Virus', *Infection, Genetics
    and Evolution*, xix (2013), p. 295; Goro Kuno, 'The Absence of Yellow Fever
    in Asia: History, Hypotheses, Vector Dispersal, Possibility of yf in Asia, and
    Other Enigmas', *Viruses*, xii/12 (2020), p. 1349; Uli Beisel and Christophe
    Boëte, 'The Flying Public Health Tool: Genetically Modified Mosquitoes
    and Malaria Control', *Science as Culture*, xxii/1 (2013), pp. 38–60.
26 Francesco Severini et al., 'Vector Competence of Italian *Aedes albopictus*
    Populations for the Chikungunya Virus (e1–226v)', *PLOS Neglected Tropical
    Diseases*, xii/4 (2018), p. e0006435; Pietro Serra, 'Con l'estate tornano
    le zanzare, I consigli per liberarsene definitivamente', *GalluraOggi.it*,
    www.galluraoggi.it, 26 June 2021; A. Marchi and L. Munstermann, 'The
    Mosquitoes of Sardinia: Species Records 35 Years After the Malaria
    Eradication Campaign', *Medical Veterinary Entomology*, I/1 (1987), pp. 89–26,
    doi: 10.1111/j.1365-2915.1987.tb00327.x; Cipriano Foxi et al., 'Entomological
    Surveillance of Zika Virus in Sardinia, Italy, 2016', *Veterinaria Italiana*, LIV/3
    (2018), doi: 10.12834/VetIt.1303.7208.2.
27 Ferran Navarro and Maite Muniesa, 'Phages in the Human Body', *Frontiers
    in Microbiology*, viii/566 (2017), doi: 10.3389/fmicb.2017.00566.
28 Louis-Marie Bobay, Marie Touchon and Eduardo Rocha, 'Pervasive
    Domestication of Defective Prophages by Bacteria', *PNAS*, cxi/33 (2014),
    pp. 12127–32, doi: 10.1073/pnas.14053361; Benjamin Chan et al., 'Phage
    Selection Restores Antibiotic Sensitivity in MDR *Pseudomonas aeruginosa*',
    Nature Scientific Reports, vi/1 (2016), art. 26717, doi: 10.1038/srep26717.
29 Richard Dawkins, *The Selfish Gene* [1976] (Oxford, 1989), p. 238.
30 Philip Madgwick, Lawrence Belcher and Jason Wolf, 'Greenbeard Genes:
    Theory and Reality', *Trends in Ecology and Evolution*, xxxiv/12 (2019),
    doi: 10.1016/j.tree.2019.08.001.
31 Arwa Elaagip, Sabrina Absalon and Anat Florentin, 'Apicoplast Dynamics
    During Plasmodium Cell Cycle', *Frontiers in Cellular and Infection
    Microbiology*, xii (2022), doi: 10.3389/fcimb.2022.864819; Avinaba Mukherjee
    and Gobinda Sadhukhan, 'Anti-Malarial Drug Design by Targeting
    Apicoplasts: New Perspectives', *Journal of Pharmacopuncture*, xix/1 (2016),
    pp. 7–15, doi: 10.3831/KPI.2016.19.001. 'When we try to pick out anything by
    itself, we find it hitched to everything else in the universe': quote from John
    Muir, *My First Summer in the Sierra* (Boston, MA, 1911), p. 211.
32 Eugeni Cuesta et al., 'Comprehensive Ecological and Geographic
    Characterization of Eukaryotic and Prokaryotic Microbiomes in
    African Anopheles', *Frontiers in Microbiology*, xii (2021), doi: 10.3389/
    fmicb.2021.635772.
33 El Hadji Amadou Niang et al., 'Biological Control of Mosquito-Borne
    Diseases: The Potential of *Wolbachia*-Based Interventions in an IVM
    Framework', *Journal of Tropical Medicine*, I (2018), doi: 10.1155/2018/
    1470459; P. Sirisena, A. Kumar and S. Sunil, 'Evaluation of *Aedes aegypti*

(Diptera: Culicidae) Life Table Attributes Upon Chikungunya Virus Replication Reveals Impact on Egg-Laying Pathways', *Journal of Medical Entomology*, LV/6 (2018), pp. 1580–87, doi: 10.1093/jme/tjy097; Sandy Ong, 'Wolbachia Goes to Work in the War on Mosquitoes', *Nature*, DXCVIII (2021), pp. S32–4, doi: 10.1038/d41586-021-02914-8; J. Denton et al., 'International Shipments of Wolbachia-Infected Mosquito Eggs – Towards Scale-up of World Mosquito Program Operations', *Revue scientifique et technique (International Office of Epizootics)*, XLI/1 (2022), pp. 91–9.

34 Nigel Turvey, *Cane Toads: A Tale of Sugar, Politics, and Flawed Science* (Sydney, 2013); Buchori Damayanti et al., *Risk Assessment on the Release of Wolbachia-Infected Aedes aegypti* (Jakarta, 2017), p. 62.

35 Jan Sapp, *Evolution by Association: A History of Symbiosis* (New York, 1994), pp. 51, 77; Susannah Porter, 'Insights into Eukaryogenesis from the Fossil Record', *Interface Focus*, X/4 (2020), doi: 10.1098/rsfs.2019.0105; Roman Zug and Peter Hammerstein, 'Bad Guys Turned Nice? A Critical Assessment of *Wolbachia* Mutualism in Arthropod Hosts', *Biological Reviews*, XC/1 (2015), pp. 89–111, doi: 10.1111/brv.12098.

36 István Zachar and Gergely Boza, 'Endosymbiosis before Eukaryotes: Mitochondrial Establishment in Protoeukaryotes', *Cellular and Molecular Life Sciences*, LXXVII/18 (2020), pp. 3503–23; David A. Baum and Buzz Baum, 'An Inside-Out Origin for the Eukaryotic Cell', *BMC Biology*, XII/76 (2014), doi: 10.1186/s12915-014-0076-2.

37 Rebecca Meaney et al., 'Designer Endosymbionts: Converting Free-Living Bacteria Into Organelles', *Current Opinion in Systems Biology*, XXIV (2020), pp. 41–50, doi: 10.1016/j.coisb.2020.09.008.

38 Jason Cournoyer et al., 'Engineering Artificial Photosynthetic Life-Forms through Endosymbiosis', *Nature Communications*, XIII (2022), art. 2254, doi: 10.1038/s41467-022-29961-7; Kartik Puri, Vito Butardo and Huseyin Sumer, 'Evaluation of Natural Endosymbiosis for Progress Towards Artificial Endosymbiosis', *Symbiosis*, LXXXIV/1 (2021), pp. 1–17, doi: 10.1007/s13199-020-00741-5.

39 Antonino Crescenti, 'Casu marzu: il formaggio coi vermi più pregiato al mondo', *InformaCibo*, www.informacibo.it, 15 January 2020.

40 'Most Dangerous Cheese', *Guinness World Records 2009*, ed. Craig Glenday (London, 2008), p. 123; 'Casu Marzu – The Illegal Cheese That's Crawling with Maggots', www.youtube.com, 15 May 2019; Anna Tsing, *The Mushroom at the End of the World* (Princeton, NJ, 2015), p. vii.

41 Carlo Ginzburg, *Il formaggio e i vermi* [1976], trans. John and Anne Tedeschi as *The Cheese and the Worms* (New York, 1982), p. 6.

42 John Farley, 'Parasites and the Germ Theory of Disease', in *Framing Disease: Studies in Cultural History* [1992], ed. Charles Rosenberg and Janet Golden (New Brunswick, NJ, 1997), p. 34; R. Peduzzi and J. Piffaretti, '*Ancylostoma duodenale* and the Saint Gothard Anaemia', *British Medical Journal*, CCLXXVII/6409 (1983), pp. 1942–5; Edoardo Perroncito, *La Malattia dei Minatori dal S. Gottardo al Sempione. Una questione risolta* (Turin, 1909); Chris Pearson, *Dogopolis: How Dogs and Humans Made Modern New York, London, and Paris* (Chicago, IL, 2021), p. 13; Dawn Biehler, *Pests in the City: Flies, Bedbugs, Cockroaches, and Rats* (Seattle, WA, 2013).

43 Carrol Lane Fenton, Review of William Patten, *The Grand Strategy of Evolution* (1920) in *The South Atlantic Quarterly*, 21 (1922), p. 82; 'Parasites in Pigs: The Large Round Worm', *Journal of the Department of Agriculture, Victoria, Australia*, 30 (1932), p. 275; Theobald Smith, *Parasitism and Disease* (Princeton, NJ, 1934), p. 95.

44 Hans Zinsser, *Rats, Lice and History* (Boston, MA, 1934), p. 60; René Dubos, Maya Pines and the Editors of Time-Life Books, *Health and Disease* (Morristown, NJ, 1965), p. 53; René Dubos cited in Pierre-Olivier Méthot, 'Why do Parasites Harm Their Host? On the Origin and Legacy of Theobald Smith's "Law of Declining Virulence"', *History and Philosophy of the Life Sciences*, XXXIV/4 (2012), p. 586; Peter Kerr, Isabella Cattadori and June Liu, 'Next Step in the Ongoing Arms Race Between Myxoma Virus and Wild Rabbits in Australia Is a Novel Disease Phenotype', *PNAS*, CXIV/35 (2017), pp. 9397–402, doi: 10.1073/pnas.1710336114.

45 Diamond, *Guns, Germs, and Steel*, 87; Alfred Crosby, *The Columbian Exchange: Biological and Cultural Consequences of 1492* (Westport, CT, 1972), p. 37; Alfred Crosby, *Ecological Imperialism: The Biological Expansion of Europe, 900–1900* (Cambridge, 1986).

46 Diamond, *Guns, Germs, and Steel*, pp. 198–200.

47 David S. Jones, 'Virgin Soils Revisited', *William and Mary Quarterly*, LX/4 (2003), p. 730; Diana Faur et al., 'Mono-Bacterial Peritonitis Caused by *Bacteroides thetaiotaomicron* in a Patient on Peritoneal Dialysis', *Nefrologia*, XXXII/5 (2012), pp. 555–701, doi: 10.3265/Nefrologia.pre2012.Jun.11568; Michael Morowitz, Erica Carlisle and John Alverdy, 'Contributions of Intestinal Bacteria to Nutrition and Metabolism in the Critically Ill', *Surgical Clinics of North America*, XCI/4 (2011), pp. 771–85, doi: 10.1016/j.suc.2011.05.001.

48 William McNeill, 'History Upside Down', *New York Review of Books*, 15 May 1997, pp. 48–50; Jones, 'Virgin Soils Revisited', p. 705; David S. Jones, *Rationalizing Epidemics: Meaning and Uses of American Indian Mortality since 1600* (Cambridge, MA, 2004).

6 Parasites R Us

1 Alfred Stefferud, ed., *Plant Diseases: The Yearbook of Agriculture* (Washington, DC, 1953), pp. 385, 723, 726; Alessandro Passera et al., 'Not Just a Pathogen? Description of a Plant-Beneficial *Pseudomonas syringae* Strain', *Frontiers in Microbiology*, X (2019), p. 1409, doi: 10.3389/fmicb.2019.01409.

2 TEDx Talks, 'The Rainmaker Named Sue: David Sands at TEDxBozeman', www.youtube.com, 20 April 2012.

3 David Sands et al., 'The Association Between Bacteria and Rain and Possible Resultant Meteorological Implications', *Időjárás: Az Országos Meteorológiai Szolgálat folyóirata*, XIII/2–4 (1982), pp. 148–52; Cindy Morris et al., 'The Life History of the Plant Pathogen *Pseudomonas syringae* Is Linked to the Water Cycle', *ISME Journal*, II (2008), pp. 321–4, doi:10.1038/ismej.2007.113; James E. Lovelock, *Gaia: A New Look at Life on Earth* [1979] (Oxford, 1987); Lynn Margulis, *Symbiotic Planet: A New Look at Evolution* (New York, 1998).

4 Kristine Harper and Ronald Doel, 'Environmental Diplomacy in the Cold War: Weather Control, the United States, and India, 1966–67', in *Environmental Histories of the Cold War*, ed. John McNeill and Corinna Unger (Cambridge, 2010), pp. 115–38; Patrick Monahan, 'Video: These Microbes Are Key to Making Artificial Snow', *Science*, 22 April 2016, doi: 10.1126/science.aaf9965; Brent Christner et al., 'Ubiquity of Biological Ice Nucleators in Snowfall', *Science*, CCCXIX/5867 (2008), doi: 10.1126/science.1149757.

5 Michel Serres, *The Parasite* [1980], trans. Lawrence Schehr (Baltimore, MD, 1982), p. 24.

6 Henry Nash Smith, 'Rain Follows the Plow: The Notion of Increased Rainfall for the Great Plains, 1844–1880', *Huntington Library Quarterly*, X/2 (1947), pp. 169–93; n.a., *Scientific American*, 37 (24 April 1880), p. 256; Stephen Cattle, 'The Case for a Southeastern Australian Dust Bowl, 1895–1945', *Aeolian Research*, XXI (2016), pp. 1–20, doi: 10.1016/j.aeolia.2016.02.001.

7 Cindy Morris et al., 'Bioprecipitation: A Feedback Cycle Linking Earth History, Ecosystem Dynamics and Land Use through Biological Ice Nucleators in the Atmosphere', *Global Change Biology*, XX/2 (2013), pp. 341–51, doi: 10.1111/gcb.12447; Vasant Saberwal, 'Science and the Desiccationist Discourse of the 20th Century', *Environment and History*, IV/3 (1998), pp. 309–43; Bruce van Haveren, 'A Reevaluation of the Wagon Wheel Gap Forest Watershed Experiment', *Forest Science*, XXXIV/1 (1988), pp. 208–14, doi: 10.1093/forestscience/34.1.208.

8 Lovelock, *Gaia: A New Look at Life on Earth*, pp. 107–8; Margulis, *Symbiotic Planet*, p. 128; James Lovelock, *The Revenge of Gaia* (London, 2006), pp. 189, 196; Stephen Harding and Lynn Margulis, 'Water Gaia: 3.5 Thousand Million Years of Wetness on Planet Earth', in *Gaia in Turmoil: Climate Change, Biodepletion, and Earth Ethics in an Age of Crisis*, ed. Eileen Crist and Bruce Rinker (Cambridge, MA, 2009), p. 55.

9 Lynn Margulis, 'Letters: Gaia in *Science*', *Science*, XXLIX/5096 (February 1993), p. 745; Lynn Margulis and Dorion Sagan, *What Is Life?* (Berkeley, CA, 1995), p. 31.

10 James Lovelock, *Healing Gaia: Practical Medicine for the Planet* (New York, 1991), pp. 12, 16.

11 James Lovelock to Lynn Margulis, 5 January 2002, in *Writing Gaia: The Scientific Correspondence of James Lovelock and Lynn Margulis*, ed. Bruce Clarke and Sébastien Dutreuil (Cambridge, 2022), p. 372; Vladimir I. Vernadsky, *The Biosphere* [1926], trans. Mark McMenamin and David Langmuir (New York, 1998), p. 14; see also letter about 'The Biosphere by Vladimir Vernadsky; an abridged version based on the French edition of 1929, Synergetic Press, Inc.', n.d., pp. 1, 3, in Margulis Folder, Box 20, James Lovelock Papers, Wroughton Museum Archives, UK.

12 Rendy Ruvindy et al., 'Unraveling Core Microbial Metabolisms in the Hypersaline Microbial Mats of Shark Bay Using High-Throughput Metagenomics', *ISME Journal*, X (2016), pp. 183–96.

13 Lovelock, *Healing Gaia*, p. 155; Margulis and Sagan, *What Is Life?*, p. 246.

14 Paul Crutzen to J. E. Lovelock, 11 February 1974, James Lovelock to Paul J.
    Crutzen, 1 March 1982, in Box 75, Part 1, James Lovelock Papers, Wroughton
    Museum Archives, UK; see also Paul J. Crutzen and Ian E. Galbally,
    'Atmospheric Conditions after a Nuclear War', *Studies in Environmental
    Science*, XXVI (1986), pp. 457–506; James Lovelock, *A Rough Ride to the
    Future* (London, 2014); Margulis, *Symbiotic Planet*, p. 115.

15 James Lovelock, 'The Fallible Concept of Stewardship of the Earth',
    6 September 2000, lecture delivered at St Georges House, Windsor, UK, in
    Box 27, Folder 9, James Lovelock Papers, Wroughton Museum Archives, UK.

16 James W. Kircher, 'The Gaia Hypothesis: Can It Be Tested?', *Reviews of
    Geophysics*, XXVII/2 (1989), pp. 223–35, doi: 10.1029/RG027i002p00223;
    George E. Brown and Anthony Ellsworth Scoville, 'The Greenhouse
    Civilization and the Gaia Hypothesis: A View from Congress' and Daniel
    A. Lashof, 'Gaia on the Brink: Biogeochemical Feedback Processes in Global
    Warming', in *Scientists on Gaia*, ed. Stephen H. Schneider and Penelope J.
    Boston (Cambridge, MA, 1993), pp. 33, 400, 409.

17 Vladimir Vernadsky quoted in Kendall Bailes, *Science and Russian Culture
    in an Age of Revolutions: V. I. Vernadsky and His Scientific School, 1863–1945*
    (Bloomington, IN, 1990), p. 194.

18 James Lovelock, *Novacene: The Coming Age of Hyperintelligence* (Cambridge,
    MA, 2019), p. 28.

19 See Philipp Lehmann, *Desert Edens: Colonial Climate Engineering in the Age
    of Anxiety* (Princeton, NJ, 2022), pp. 38–52; Daniel O'Neill, *Firecracker Boys:
    H-Bombs, Inupiat Eskimos, and the Roots of the Environmental Movement*
    (New York, 1994); Milo Nordyke, 'The Soviet Program for Peaceful Uses
    of Nuclear Explosions' (1996), at Lawrence Livermore National Laboratory,
    UCRL-ID-124410, https://inis.iaea.org, accessed 1 April 2024; Katja Doose,
    'A Global Problem in a Divided World: Climate Change Research During
    the Late Cold War, 1972–1991', *Cold War History*, XXI/4 (2021), pp. 469–89.

20 James Lovelock and Sidney Epton, 'The Quest for Gaia', *New Scientist*
    LXV/935 (1975), p. 306; Erle Ellis et al., 'People Have Shaped Most of
    Terrestrial Nature for at Least 12,000 Years', *PNAS*, CXVIII/17 (2021),
    p. e2023483118; Richard Lewontin, 'Gene, Organism and Environment', in
    Evolution from Molecules to Men, ed. D. S. Bendell (Cambridge, 1983), p. 280.

21 Kevin Laland, John Odling-Smee and John Endler, 'Niche Construction,
    Sources of Selection and Trait Coevolution', *Interface Focus*, VII/5 (2017),
    doi: 10.1098/rsfs.2016.0147.

22 Guilherme F. De Araújo, Renan C. Moioli and Sandro J. de Souza, 'The
    Shared Use of Extended Phenotypes Increase the Fitness of Simulated
    Populations', *Frontiers in Genetics*, XII (March 2021), doi: 10.3389/
    fgene.2021.617915.

23 Bruce Smith and Melinda Zeder, 'The Onset of the Anthropocene',
    Anthropocene, IV (2013), pp. 8–13.

24 Nicole Boivin, 'Human and Human-Mediated Species Dispersals through
    Time: Introduction and Overview', in *Human Dispersal and Species
    Movement: From Prehistory to the Present*, ed. Nicole Boivin, Rémy Crassard
    and Michael Petraglia (Cambridge, 2017), pp. 3–26.

25 Margulis, *What Is Life?*, pp. 131–2.

26 Lynn Margulis, ed., *Proceedings of the Second Conference on Origins of Life: Cosmic Evolution, Abundance, and Distribution of Biologically Important Elements* (New York, 1971), pp. 106, 159.

27 Margulis, *Symbiotic Planet*, p. 128; Carl Sagan, Transcript from 'Cosmos: A Personal Voyage', PBS Series, Part 13 (1980).

28 Daniel Pauly and Dirk Zeller, 'Catch Reconstructions Reveal that Global Marine Fisheries Catches Are Higher Than Reported and Declining', *Nature Communications*, VII (2016), doi: 10.1038/ncomms10244; M.L.D. Palomares et al., 'Fishery Biomass Trends of Exploited Fish Populations in Marine Ecoregions, Climatic Zones and Ocean Basins', *Estuarine, Coastal and Shelf Science*, 243 (2020), doi: 10.1016/j.ecss.2020.106896.

29 Paul Ehrlich, *The Population Bomb* (New York, 1968), p. xi; Scott Russell Sanders, *A Conservationist Manifesto* (Bloomington, IN, 2009), p. 47; Warren M. Hern, 'Has the Human Species Become a Cancer on the Planet? A Theoretical View of Population Growth as a Sign of Pathology', *Current World Leaders*, XXXVI/6 (1993), abstract.

30 David R. Brower with Steve Chapple, *Let the Mountains Talk, Let the Rivers Run: A Call to Those Who Would Save the Earth* (San Francisco, CA, 1995), p. 50; Louise Gray, 'David Attenborough – Humans Are Plague on Earth', *The Telegraph*, 22 January 2013; Donald Worster, 'Another Silent Spring', *Environment and Society Portal*, Virtual Exhibitions 2020, no. 1 (22 April 2020), Rachel Carson Center for Environment and Society, doi: 10.5282/rcc/9028.

31 Lovelock, *Healing Gaia*, pp. 154, 171.

32 Hans Zinsser, *Rats, Lice and History* (London, 1935), p. 9.

33 Rachel Carson, *Silent Spring* (New York, 1962), p.99; René Dubos, 'Symbiosis between the Earth and Humankind', *Science*, CXCIII/4252 (1976), pp. 459–62.

34 Georges Louis Leclerc, comte de Buffon, *Histoire Naturelle* (Paris, 1749), trans. C. D. Warner et al., in The Library of the World's Best Literature, www.bartleby.com/library/prose/1008.html, accessed 1 August 2024.

35 Genesis 2:15; Sahih Muslim, hadith no. 2742; Georges Louis Leclerc, comte de Buffon, quoted in Lehmann, *Desert Edens*, 8.

36 Divya Sehgal, 'What Animal Collectively Makes Up the Largest Biomass on Earth?', www.sciencefocus.com, accessed 28 June 2022; Hafiz Ishfaq Ahmad et al., 'The Domestication Makeup: Evolution, Survival, and Challenges', *Frontiers in Ecology and Evolution*, VIII (2020), doi: 10.3389/fevo.2020.00103; USDA, 'United States and Canadian Cattle and Sheep' (2022), www.nass.usda.gov, accessed 1 August 2024; U.S. National Park Service, 'Frequently Asked Question: Bison' (2021), www.nps.gov, accessed 28 June 2022; Rob Cook, 'Ranking of Countries with the Most Cattle', https://beef2live.com, accessed 5 August 2023; Yinon M. Bar-On, Rob Phillips and Ron Milo, 'The Biomass Distribution on Earth', *Proceedings of the National Academy of Sciences*, CXV/25 (2018), pp. 6506–11, doi: 10.1073/pnas.1711842115; Jianping Huang et al., 'The Global Oxygen Budget and Its Future Projection', *Science Bulletin*, LXIII/18 (2018), pp. 1180–86.

37 Bar-On, Phillips and Milo, 'The Biomass Distribution on Earth'; Hannah Ritchie, 'The World Has Lost One-Third of Its Forest, But an End of Deforestation Is Possible', https://ourworldindata.org, 9 February 2021;

George Perkins Marsh, *Man and Nature; or, Physical Nature as Modified by Human Action* (New York, 1864), p. 36; Vernadsky, *The Biosphere*, p. 151.

38 Marsh, *Man and Nature*, p. 14.

39 Pierre Joseph van Beneden, *Animal Parasites and Messmates*, 2nd edn (London, 1876), p. 115.

40 Davin, 'World's Oldest Terrarium/Sealed Bottle Ecosystem by David Latimer – Only Watered Once in 53 Years', https://biologicperformance.com, 26 March 2022.

41 Luke Keogh, 'How a Simple Box Moved the Plant Kingdom', *Arnoldia*, LXXIV/4 (2017), pp. 2–13; Christopher Thacker, *The History of Gardens* (Berkeley, CA, 1979), p. 236.

42 Frank Salisbury, Josef Gitelson and Genry Lisovsky, 'Bios-3: Siberian Experiments in Bioregenerative Life Support', *Bioscience*, XLVII/9 (1997), pp. 575–84.

43 John Petersen et al., 'The Making of Biosphere 2', *Restoration and Management Notes*, X/2 (1992), pp. 158–68; Margaret Augustine, 'Presentation to National Commission on Space', in Carton 24, Folder 3, Frank Salisbury Papers, University of Utah; John Allen quoted in Mark Nelson, 'Summary of Meeting on History, Current Work and Future of Closed Ecological Systems', [n.d. but *c.* 1988] in Carton 64, Folder 3, Frank Salisbury Papers, University of Utah. On Biosphere 2's conception and development, see also Peder Anker, 'The Ecological Colonization of Space', *Environmental History*, X/2 (2005), pp. 239–68.

44 Lisa Ruth Rand, 'The Glass Ark, and Other Fraught Analogies', in 'Biosphere 2: Why an Eccentric Ecological Experiment Still Matters 25 Years Later', *Edge Effects*, 15 December 2016, updated 9 October 2021, at https://edgeeffects.net; Luc Peters, *On Noise! Philosophy – Art – Organization* (Newcastle upon Tyne, 2000), p. 38.

45 John Allen and Mark Nelson quoted in Rebecca Reider, *Dreaming the Biosphere* (Tucson, AZ, 2009), p. 121; Eugene Odum, *Ecology* (New York, 1963), p. 109; Letter from Abigail 'Gaie' Alling and Mark Nelson to Frank Salisbury and Bruce Bugbee, 31 May 2024, in Carton 64, Folder 3, Frank Salisbury Papers, University of Utah; James Lovelock, 'Foreword' in Fred Pearce, *Earth: Then and Now* (Buffalo, NY, 2010), p. 9.

46 Keridwen Cornelius, 'Biosphere 2: The Once Infamous Live-In Terrarium is Transforming Climate Research', *Scientific American*, www.scientificamerican.com, 4 October 2021.

47 Angus Wright, 'Review of *Dreaming the Biosphere* by Rebecca Reider', *Agricultural History*, LXXXIV/4 (2010), pp. 551–2.

7 Restoring Parasites and Rewilding Microbiomes

1 Jong-Shik Kim et al., 'Bacterial Diversity of Terra Preta and Pristine Forest Soil from the Western Amazon', *Soil Biology and Biochemistry*, XXXIX/2 (2007), pp. 684–90; William Demetrio et al., 'A "Dirty" Footprint: Macroinvertebrate Diversity in Amazonian Anthropic Soils', *Global Change Biology*, XXVII/19 (2021), pp. 4575–91; Muhammad Nadeem et al., 'Soil Microbes and Climate-Smart Agriculture', in *Global Agricultural Production:*

Resilience to Climate Change, ed. M. Ahmed (Springer, 2022), pp. 107–47; Terra Preta, https://terrapreta.ca, accessed 21 May 2024; James Lovelock, *Novacene: The Coming Age of Hyperintelligence* (Cambridge, MA, 2019).

2 'Pleistocene Park', https://pleistocenepark.ru, accessed 22 May 2024; Igor Popov, 'The Current State of Pleistocene Park, Russia (An Experiment in the Restoration of Megafauna in a Boreal Environment', *The Holocene*, XXX/10 (2020), doi: 10.1177/0959683620932975.

3 Maria Gloria Dominguez-Bello quoted in 'Can We Save Our Body's Ecosystem from Extinction?', PBS News Hour, www.pbs.org, 23 April 2014; Pajau Vangay et al., 'U.S. Immigration Westernizes the Human Gut Microbiome', *Cell*, CLXXV/4 (2018), pp. 962–72.e10; Francesco Morandini et al., 'Urbanization Associates with Restricted Gut Microbiome Diversity and Delayed Maturation in Infants', *iScience*, XXVI/11 (2023), 108136; Christopher Wallen-Russell et al., 'A Catastrophic Biodiversity Loss in the Environment Is Being Replicated on the Skin Microbiome: Is This a Major Contributor to the Chronic Disease Epidemic?', *Microorganisms*, XI/11 (2023), 2784; Nicola Segata, 'Gut Microbiome: Westernization and the Disappearance of Intestinal Diversity', *Current Biology*, XXV/14 (2015), pp. R611–13.

4 Colin Carlson et al., 'Parasite Diversity Faces Extinction and Redistribution in a Changing Climate', *Science*, III/9 (2017), doi: 10.1126/sciadv.160242; Robert Dunn et al., 'The Sixth Mass Coextinction: Are Most Endangered Species Parasites and Mutualists?', *Proceedings of the Royal Society B*, 276 (2009), pp. 3037–45; Chelsea Wood et al., 'A Reconstruction of Parasite Burden Reveals One Century of Climate-Associated Parasite Decline', PNAS, CXX/3 (2023), doi: 10.1073/pnas.2211903120.

5 Joshua Brian, 'Parasites in Biodiversity Conservation: Friend Or Foe?', *Trends in Parasitology*, XXXIX/8 (2023), doi: 10.1016/j.pt.2023.05.005; Shamik Dholakia et al., 'Pubic Lice: An Endangered Species?', *Sexually Transmitted Diseases*, XLI/6 (2014), pp. 388–91; Nadia Amanzougaghene et al., 'Where Are We With Human Lice? A Review of the Current State of Knowledge', *Frontiers in Cellular and Infection Microbiology*, IX (2019), doi: 10.3389/fcimb.2019.00474

6 Allan Radaic and Yvonne Kapila, 'The Oralome and Its Dysbiosis: New Insights into Oral Microbiome-Host Interactions', *Computational and Structural Biotechnology Journal*, XIX (2021), pp. 1335–60; M. Addy and M. Hunter, 'Can Tooth Brushing Damage Your Health? Effects on Oral and Dental Tissues', *International Dental Journal*, LIII (2003), pp. 177–86; 'Is Over Brushing Your Teeth Bad?' (2023), www.colgate.com, accessed 1 September 2024.

7 Radaic and Kapila, 'The Oralome and Its Dysbiosis'; Trang Nguyen et al., 'Probiotics for Periodontal Health: Current Molecular Findings', *Periodontology*, LXXXVII/1 (2021), doi: 10.1111/prd.12382.

8 Barbara Moroni et al., 'Unintentional Recovery of Parasitic Diversity Following Restoration of Red Deer (*Cervus elaphus*) in North-Western Italy', *Animals*, XII/11 (2022), doi: 10.3390/ani12111433; Colin Carlson et al., 'A Global Parasite Conservation Plan', *Biological Conservation*, CCL (2020), 108596; Cock van Oosterhout et al., 'The Guppy as a Conservation Model:

Implications of Parasitism and Inbreeding for Reintroduction Success',
Conservation Biology, XXI/6 (2007), doi: 10.1111/j.1523-1739.2007.00809.x.

9 Catherine Burns, Brett Goodwin and Richard Ostfeld, 'A Prescription for
Longer Life? Bot Fly Parasitism of the White-Footed Mouse', *Ecology*,
LXXXVI/3 (2005), pp. 753–61; John Vucetich, *Restoring the Balance: What
Wolves Tell Us about Our Relationship with Nature* (Baltimore, MD, 2021),
pp. 45–6.

10 J. Gerlach, '*Haematopinus oliveri*', *IUCN Red List of Threatened Species* (2014),
e.T9621A21423551; Bret Boyd et al., 'Primates, Lice and Bacteria: Speciation
and Genome Evolution in the Symbionts of Hominid Lice', *Molecular
Biology and Evolution*, XXXIV/7 (2017), doi: 10.1093/molbev/msx117.

11 See 'Svalbard Global Seed Vault', www.frozenark.org; 'Biorepository',
https://naturalhistory.si.edu/research/biorepository, both accessed
1 September 2024; Kristin O'Brien et al., 'The Time Is Right for an
Antarctic Biorepository Network', *PNAS*, CXIX/50 (2022), p. e2212800119.

12 'Probiotics: What You Need To Know', www.nccih.nih.gov; 'Food: Probiotic
Claims', www.asa.org.uk; www.yakult.co.uk, all accessed 31 May 2024.

13 J. Hamilton-Miller, G. Gibson and W. Bruck, 'Some Insights Into the
Derivation and Early Uses of the Word "Probiotic"', *British Journal of
Nutrition*, XC/4 (2003), p. 845; Sasiprapa Krongdang et al., 'Edible Insects
in Thailand: An Overview of Status, Properties, Processing, and Utilization in
the Food Industry', *Foods*, XII/11 (2023), p. 2162; Valerie Stull et al., 'Impact of
Edible Cricket Consumption on Gut Microbiota in Healthy Adults, a Double-
Blind, Randomized Crossover Trial', *Scientific Reports*, VIII/1 (2018), 10762;
Giovanni Gasbarrini, Fiorenza Bonvicini and Annagiulia Gramenzi,
'Probiotics History', *Journal of Clinical Gastroenterology*, L (2016), pp. S116–19;
Lynne McFarland, 'From Yaks to Yogurt: The History, Development, and
Current Use of Probiotics', *Clinical Infectious Diseases*, LX/Suppl 2 (2015),
pp. S85–90; Reinhard Hoeppli and I-Hung Ch'iang, 'Parasites in Chinese
and Early Western Medicine: A Comparison', *Peking Natural History Bulletin*,
XVIII/4 (1950), pp. 207–43.

14 Gretel Pelto, Yuanyuan Zhang and Jean-Pierre Habicht, 'Premastication:
The Second Arm of Infant and Young Child Feeding for Health and
Survival?', *Maternal and Child Nutrition*, VI/1 (2010), pp. 4–18; Penny
Van Esterik et al., 'Commentaries on "Premastication" by G. Pelto et al.',
Maternal and Child Nutrition, VI/1 (2010), pp. 19–26; Khalifa Sharquie,
Fatema Al-Jaralla and Inas Sharquie, 'The Human Microbiomes as a
Dynamic Process: Old and New Ideas', *Journal of Pakistan Association of
Dermatologists*, XXXII/2 (2022), pp. 396–405.

15 Adrian Boicean et al., 'Fecal Microbiota Transplantation in Inflammatory
Bowel Disease', *Biomedicines*, XI/4 (2023), 1016; Faming Zhang et al., 'Should
We Standardize the 1,700-Year-Old Fecal Microbiota Transplantation?',
American Journal of Gastroenterology, CVII/11 (2012), doi: 10.1038/ajg.2012.251.

16 Petra Maurer, 'Faeces and the Old Sole of a Shoe: Remedies of the
Dreckapotheke', *ASIA*, LXXI/4 (2017), pp. 1247–92, doi: 10.1515/asia-2017-0050;
John Hunter, 'Geophagy in Africa and the United States: A Culture-Nutrition
Hypothesis', *Geographical Review*, LXIII/2 (1973), pp. 170–95; J. Croese et al.,
'A Proof of Concept Study Establishing *Necator Americanus* in Crohn's

Patients and Reservoir Donors', *Gut*, LV/1 (2006), pp. 136–7, doi: 10.1136/gut.2005.079129; Luisa Mourão Dias Magalhães et al., 'Immunopathology and Modulation Induced by Hookworms: From Understanding to Intervention', *Parasite Immunology*, XCIII/2 (2020), doi: 10.1111/pim.12798.

17 Gabrielle Palmer, *The Politics of Breastfeeding: When Breasts Are Bad for Business* (London, 2009).

18 Radaic and Kapila, 'The Oralome and Its Dysbiosis'; Ara Koh et al., 'From Dietary Fiber to Host Physiology: Short-Chain Fatty Acids as Key Bacterial Metabolites', *Cell*, CLXV/6 (2016), pp. 1332–45.

19 K. Arun, P. Nisha and Alberto Finamore, 'Editorial: Emerging Perspectives on Probiotics, Prebiotics, and Synbiotics for Prevention and Management of Chronic Disease', *Frontiers in Nutrition*, X (2023), doi: 10.3389/fnut.2023.1309140; Se Jin Song et al., 'Cohabiting Family Members Share Microbiota with One Another and with Their Dogs', *eLife*, XVI/2 (2013), doi: 10.7554/eLife.00458.

20 Gabriel Vinderola, Mary Ellen Sanders and Seppo Salminen, 'The Concept of Postbiotics', *Foods*, XI/8 (2022), p. 1077; Ernest Faust, *Human Helminthology* (Philadelphia, PA, 1929), p. 2.

21 Jamie Lorimer, *The Probiotic Planet: Using Life to Manage Life* (Minneapolis, MN, 2020), pp. 137–59.

22 Frans Vera, 'The Shifting Baseline Syndrome in Restoration Ecology', in *Restoration and History: The Search for a Usable Environmental Past*, ed. Marcus Hall (New York, 2010), pp. 98–110.

23 George Monbiot, *Feral: Rewilding the Land, The Sea and Human Life* (Toronto, CA, 2013), pp. 84–6; Samiran Banerjee, Klaus Schlaeppi and Marcel van der Heijden, 'Keystone Taxa as Drivers of Microbiome Structure and Functioning', *Nature Reviews Microbiology*, XVI/9 (2018), pp. 567–76; Warder Allee et al., *Principles of Animal Ecology* (Philadelphia, PA, 1949), p. 254; Carl Zimmer also made this claim about the third habitat in his *Parasite Rex: Inside the Bizarre World of Nature's Most Dangerous Creatures* (New York, 2000), p. 156.

24 Marcus Hall, 'Extracting Culture or Injecting Nature? Rewilding in Transatlantic Perspective', in *Old World and New World Perspectives on Environmental Philosophy,* ed. Jozef Keulartz and Martin Drenthen (New York, 2014), pp. 17–35; Jamie Lorimer and Clemens Driessen, 'From "Nazi Cows" to Cosmopolitan "Ecological Engineers": Specifying Rewilding through a History of Heck Cattle', *Annals of the American Association of Geographers*, CVI/3 (2016), pp. 631–52; Marcus Hall, 'The High Art of Rewilding: Lessons from Curating Earth Art', in *Rewilding*, ed. N. Pettorelli, S. Durant and J. du Toit (Cambridge, 2019), pp. 201–21.

25 Aisha Williams, 'The Absurdity of Rewilding: A Philosophical Take on Conservation (2022), www.meer.com, accessed 1 September 2024; Eric Higgs et al., 'The Changing Role of History in Restoration Ecology', *Frontiers in Ecology and Environment*, XII/9 (2014), pp. 499–506.

26 *Agriculture International*, vol. XLI–XLII (Ithaca, NY, 1990), p. 20; Rob Dunn, *The Wild Life of Our Bodies: Predators, Parasites, and Partners that Shape Who We Are Today* (New York, 2011); Pauline Doligez et al., 'Promoting Grass in Horse Diets and Implementing Sustainable Deworming', in *Equine Science,*

ed. C. Rutland and A Rizvanov (London, 2020), pp. 185–29; Aline Hoffmann et al., 'The Skin Microbiome in Healthy and Allergic Dogs', PLOS ONE, IX/1 (2014), p. e83197.

27 Hossain Md Anawar and Rezaul Chowdhury, 'Remediation of Polluted River Water by Biological, Chemical, Ecological and Engineering Processes', *Sustainability*, XII/7 (2020), 7017.

28 Isabel Reche et al., 'Deposition Rates of Viruses and Bacteria Above the Atmospheric Boundary Layer', *ISME Journal*, XII/4 (2018), pp. 1154–62.

29 Eric P. Hoberg, 'Foundations for an Integrative Parasitology: Collections, Archives, and Biodiversity Informatics', *Comparative Parasitology*, LXIX/2 (2002), p. 127; K. Ekbom, 'The Pre-Senile Delusion of Infestation', *History of Psychiatry*, XIV/2 (2003), trans. of K. Ekbom, 'Der praesenile Dermatozoenwahn', *Acta Psychiatrica et Neurologica Scandinavica*, 13 (1938), pp. 227–59, doi: 10.1177/0957154X03014; Beaver quoted in 'Transcripts of 1980 Conference; Current Status and Future of Parasitology', in Box E, Folder 'History of Definition of Parasitology', John S. Andrews Papers, Special Collections, U.S. National Agricultural Library.

30 'Russian "Disinformation" Hyped Paris Bedbug Scare, French Minister Says', www.reuters.com, 1 March 2024; 'France Hits Back at Hysteria Over Bedbug "Invasion"', www.france24.com, accessed 4 October 2024; Dawn Biehler, *Pests in the City: Flies, Bedbugs, Cockroaches, and Rats* (Seattle, WA, 2013), p. 207.

31 Zimmer, *Parasite Rex*; Rosemary Drisdelle, *Parasites: Tales of Humanity's Most Unwelcome Guests* (Berkeley, CA, 2010); J. G. LeBlanc et al., 'Bacteria as Vitamin Suppliers to Their Host: A Gut Microbiota Perspective', *Current Opinion in Biotechnology*, XXIV/2 (2013), pp. 160–68; Nancy Tomes, *The Gospel of Germs: Men, Women, and the Microbe in American Life* (Cambridge, MA, 1998); Leonard Sedgwick, 'Report on the Parasitic Theory of Disease', *Transactions of the St. Andrews Medical Graduates' Association* (1869), pp. 116–48; Henry Gradle quoted in Tomes, *The Gospel of Germs*, p. 44.

32 Edward O. Wilson, *Half-Earth: Our Planet's Fight for Life* (New York, 2016), p. 53.

33 Steve Paulson, 'Why a Famous Biologist Wants to Eradicate Killer Mosquitoes', https://theworld.org, 4 April 2016.

34 On the debate over whether or not we should try to eradicate mosquitoes, see Marcus Hall and Dan Tamïr, eds, *Mosquitopia: The Place of Pests in a Healthy World* (New York, 2021).

35 James McGrigor, *Medical Sketches of the Expedition to Egypt from India* (London, 1804), p. 213.

36 Magali Lemaitre et al., 'Coinfection with *Plasmodium falciparum* and *Schistosoma haematobium*: Additional Evidence of the Protective Effect of Schistosomiasis on Malaria in Senegalese Children', *American Journal of Tropical Medicine and Hygiene*, XC/2 (2014), pp. 329–34; J. Murray et al., 'The Biological Suppression of Malaria: An Ecological and Nutritional Interrelationship of a Host and Two Parasites', *American Journal of Clinical Nutrition*, 31 (1978), pp. 1363–6; Jonathan Roberts, 'Participating in Eradication: How Guinea Worm Redefined Eradication, and Eradication Redefined Guinea Worm, 1985–2022', *Medical History*, LXVII/2 (2023), pp. 148–71.

SELECT
BIBLIOGRAPHY

Ashford, Richard, and William Crewe, *The Parasites of Homo Sapiens:*
   An Annotated Checklist of the Protozoa, Helminths and Arthropods for
   Which We Are Home (London, 2003)

Blaser, Martin, *Missing Microbes: How the Overuse of Antibiotics Is Fueling*
   Our Modern Plagues (New York, 2014)

Bonelli, Franco, 'La malaria nella storia demografica ed economica
   d'Italia: Primi lineamenti di una ricerca', *Studi Storici*, VII/4 (1966),
   pp. 659–87

Bremermann, Hans, and John Pickering, 'A Game-Theoretical Model of
   Parasite Virulence', *Journal of Theoretical Biology*, C/3 (1983), pp. 411–26

Brown, Peter J., 'Microparasites and Macroparasites', *Cultural Anthropology*, II/1
   (1987), pp. 155–71

Buettner, Dan, *The Blue Zones: Lessons for Living Longer from the People Who've*
   Lived the Longest (Washington, DC, 2008)

Burns, Catherine, Brett Goodwin and Richard Ostfeld, 'A Prescription for Longer
   Life? Bot Fly Parasitism of the White-Footed Mouse', *Ecology*, LXXXVI/3
   (2005), pp. 753–61

Carlson, Colin, et al., 'A Global Parasite Conservation Plan', *Biological*
   Conservation, CCL (2020), 108596

Casadevall, Arturo, and Liise-anne Pirofski, 'Microbiology: Ditch the Term
   Pathogen', *Nature*, DXVI (2014), pp. 165–6

Crosby, Alfred, *The Columbian Exchange: Biological and Cultural Consequences of*
   1492 (Westport, CT, 1972)

Curtin, Philip, 'The End of the "White Man's Grave"? Nineteenth-Century
   Mortality in West Africa', *Journal of Interdisciplinary History*, XXI/1 (1990),
   pp. 63–88

Desowitz, Robert, *New Guinea Tapeworms and Jewish Grandmothers: Tales of*
   Parasites and People (New York, 1981)

Diamond, Jared, *Guns, Germs, and Steel: The Fates of Human Societies* (New York,
   1997)

Dubos, René, *Mirage of Health: Utopias, Progress, and Biological Change* (New
   York, 1959)

Dunn, Rob, *The Wild Life of Our Bodies: Predators, Parasites, and Partners That*
   Shape Who We Are Today (New York, 2011)

—, et al., 'The Sixth Coextinction: Are Most Endangered Species Parasites and Mutualists?', *Proceedings of the Royal Society B*, CCLXXVI/1670 (2009), pp. 3037–45

Gallego-Delgado, Julio, and Ana Rodriguez, 'Malaria and Hypertension: Another Co-Evolutionary Adaptation?', *Frontiers in Cellular and Infection Microbiology*, IV/121 (2014), doi: 10.3389/fcimb.2014.00121

Gasbarrini, Giovanni, Fiorenza Bonvicini and Annagiulia Gramenzi, 'Probiotics History', *Journal of Clinical Gastroenterology*, L (2016), pp. S116–19

Grove, David, *A History of Human Helminthology* (Wallingford, 1990)

Gullestad, Anders, 'Parasite', *Political Concepts: A Critical Lexicon*, www.politicalconcepts.org, accessed 19 July 2013

Hackett, Lewis, *Malaria in Europe: An Ecological Study* (London, 1937)

Hall, Marcus, 'Environmental Imperialism in Sardinia: Pesticides and Politics in the Struggle Against Malaria', in *Nature and History in Modern Italy*, ed. Marco Armiero and Marcus Hall (Athens, OH, 2010), pp. 70–86

—, 'The High Art of Rewilding: Lessons from Curating Earth Art', in *Rewilding*, ed. N. Pettorelli, S. Durant and J. du Toit (Cambridge, 2019), pp. 201–21

—, 'World War II and the Axis of Disease', in *War and the Environment: Military Destruction in the Modern Age,* ed. Charles Closmann (College Station, TX, 2009), pp. 112–31

—, and Dan Tamïr, eds, *Mosquitopia: The Place of Pests in a Healthy World* (London, 2021)

Harrison, Gordon, *Mosquitoes, Malaria, and Man: A History of the Hostilities since 1880* (New York, 1968)

Heatherington, Tracey, *Wild Sardinia: Indigeneity and the Global Dreamtimes of Environmentalism* (Seattle, WA, 2010)

Hoberg, Eric P., 'Foundations for an Integrative Parasitology: Collections, Archives, and Biodiversity Informatics', *Comparative Parasitology*, LXIX/2 (2002), pp. 124–31

Hoeppli, Reinhard, and I-Hung Ch'iang, 'Parasites in Chinese and Early Western Medicine: A Comparison', *Peking Natural History Bulletin*, XVIII/4 (1950), pp. 207–43

—, 'Selections from Old Chinese Medical Literature on Various Subjects of Helminthological Interest', *Chinese Medical Journal*, LVII/4 (1940), pp. 373–87

Honigsbaum, Mark, *The Fever Trail: In Search of the Cure for Malaria* (New York, 2001)

Hunter, John, 'Geophagy in Africa and the United States: A Culture-Nutrition Hypothesis', *Geographical Review*, LXIII/2 (1973), pp. 170–95

Jones, David S., *Rationalizing Epidemics: Meaning and Uses of American Indian Mortality since 1600* (Cambridge, MA, 2004)

Kropotkin, Peter, *Mutual Aid: A Factor of Evolution* (London, 1902)

Lawrence, D. H., *Sea and Sardinia* (New York, 1921)

Le Lannou, Maurice, *Pâtres et paysans de la Sardaigne* (Tours, 1941)

Lincicome, David, 'The Goodness of Parasitism: A New Hypothesis', in *Aspects of the Biology of Symbiosis*, ed. Thomas Cheng (Baltimore, MD, 1971), pp. 139–227

Lorimer, Jamie, *The Probiotic Planet: Using Life to Manage Life* (Minneapolis, MN, 2020)

Lovelock, James, *Gaia: A New Look at Life on Earth* (Oxford, 1979)

—, *Novacene: The Coming Age of Hyperintelligence* (Cambridge, MA, 2019)

—, *The Revenge of Gaia* (London, 2006)

Loy, Dorothy, et al., 'Out of Africa: Origins and Evolution of the Human Malaria Parasites *Plasmodium falciparum* and *Plasmodium vivax*', *International Journal for Parasitology*, XLVII/2–3 (2017), pp. 87–97

McNeill, John, 'How the Lowly Mosquito Helped America Win Independence', *Smithsonian Magazine*, 15 June 2016

—, *Mosquito Empires: Ecology and War in the Greater Caribbean, 1620–1914* (Cambridge, 2010)

McNeill, William, *Plagues and Peoples* (New York, 1976)

Margulis, Lynn, and Dorion Sagan, *Symbiotic Planet: A New Look at Evolution* (New York, 1998)

—, *What Is Life?* (Berkeley, CA, 1995)

Maurer, Petra, 'Faeces and the Old Sole of a Shoe: Remedies of the *Dreckapotheke*', *ASIA*, LXXI/4 (2017), pp. 1247–92

Méthot, Pierre-Olivier, 'Why Do Parasites Harm Their Host? On the Origin and Legacy of Theobald Smith's "Law of Declining Virulence"', *History and Philosophy of the Life Sciences*, XXXIV/4 (2012), pp. 561–601

Mitman, Gregg, and Rob Nixon, 'A Dialogue on Form, Knowledge, and Representation', in 'Minding the Gap: Working Across Disciplines in Environmental Studies', *RCC Perspectives*, II (2014), pp. 61–8

Øverli, Øyvind, and Ida Johansen, 'Kindness to the Final Host and Vice Versa: A Trend for Parasites Providing Easy Prey?' *Frontiers in Ecology and Evolution*, VII (2019), doi: 10.3389/fevo.2019.00050

Packard, Randall, *The Making of a Tropical Disease: A Short History of Malaria* (Baltimore, MD, 2010)

Pelto, Gretel, Yuanyuan Zhang and Jean-Pierre Habicht, 'Premastication: The Second Arm of Infant and Young Child Feeding for Health and Survival?', *Maternal and Child Nutrition*, VI/1 (2010), pp. 4–18

Pollan, Michael, *The Botany of Desire: A Plant's-Eye View of the World* (New York, 2001)

Radaic, Allan, and Yvonne Kapila, 'The Oralome and Its Dysbiosis: New Insights Into Oral Microbiome-Host Interactions', *Computational and Structural Biotechnology Journal*, XIX (2021), pp. 1335–60

Salisbury, Frank, Josef Gitelson and Genry Lisovsky, 'Bios-3: Siberian Experiments in Bioregenerative Life Support', *Bioscience*, XLVII/9 (1997), pp. 575–84

Sapp, Jan, *Evolution by Association: A History of Symbiosis* (New York, 1994)

Segata, Nicola, 'Gut Microbiome: Westernization and the Disappearance of Intestinal Diversity', *Current Biology*, XXV/14 (2015), pp. R611–13

Serres, Michel, *The Parasite* [1980], trans. Lawrence Schehr (Baltimore, MD, 1982)

Smith, Henry Nash, 'Rain Follows the Plow: The Notion of Increased Rainfall for the Great Plains, 1844–2880', *Huntington Library Quarterly*, X/2 (1947), pp. 169–93

Smith, Theobald, *Parasitism and Disease* (Princeton, NJ, 1934)

Sutter, Paul , '"The First Mountain to Be Removed": Yellow Fever Control and the Construction of the Panama Canal', *Environmental History*, XXI/2 (2016), pp. 250–59

Todes, Daniel, *Darwin Without Malthus: The Struggle for Existence in Russian Evolutionary Thought* (New York, 1989)

Tognotti, Eugenia, 'Malaria in Sardinia', *International Journal of Anthropology*, XIII/3–4 (1998), pp. 237–42

Van Beneden, Pierre Joseph, *Animal Parasites and Messmates*, 2nd edn (London, 1876)

Vernadsky, Vladimir I., *The Biosphere* [1926], trans. Mark McMenamin and David Langmuir (New York, 1998)

Walker, Kim, and Mark Nesbitt, *Just the Tonic: A Natural History of Tonic Water* (Kew, 2020)

Ward, Henry B., 'The Parasitic Worms of Man and the Domestic Animals', *Studies from the Zoological Laboratory*, University of Nebraska-Lincoln, Paper 9 (1895)

Warren, Kenneth, and Elizabeth Purcell, eds, *The Current Status and Future of Parasitology* (New York, 1981)

Webb, James, Jr, *Humanity's Burden: A Global History of Malaria* (Cambridge, 2009)

Willoughby, Urmi Engineer, 'Domesticated Mosquitoes: Colonization and Growth of Mosquito Habitats in North America', in *Mosquitopia: The Place of Pests in a Healthy World*, ed. Marcus Hall and Dan Tamïr (London, 2021), pp. 61–72

Wood, Chelsea, et al., 'A Reconstruction of Parasite Burden Reveals One Century of Climate-Associated Parasite Decline', PNAS, CXX/3 (2023), doi: 10.1073/pnas.2211903120

Worboys, Michael, 'The Emergence and Early Development of Parasitology', in *Parasitology: A Global Perspective*, ed. Kenneth Warren and John Bowers (New York, 1983), pp. 1–18

Zimmer, Carl, *Parasite Rex: Inside the Bizarre World of Nature's Most Dangerous Creatures* (New York, 2000)

Zinsser, Hans, *Rats, Lice and History* (London, 1935)

ACKNOWLEDGEMENTS

By tracing the roundabout ways that our maligned parasites may be doing more good than bad, I hope to convince others to ask further questions about how these and other co-organisms, small or large, are changing our lives, and us theirs. Many people and organizations have assisted in my slow and sometimes convoluted 'animal turn' that journeyed from the Anthropocene to Sardinia to malaria to mosquitoes to parasites and back again. Certainly the big drawback of making any list of acknowledgements is forgetting to mention crucial players, or else realizing later just how large some roles were without giving them adequate mention. I nonetheless offer here a handful of key institutions, to include the University of Zurich, for giving me the time and the space; the University of Wisconsin-Madison, for the formation and faith; the Rachel Carson Center for Environment and Society, for the inspiration and community; the Universities of Utah, Alaska and Bologna for the practice and perseverance. The Fulbright Commission, German Marshall Fund, Gladys Krieble Delmas Foundation, European University Institute, Carnegie Council for Ethics in International Affairs, American Academy of Rome and Council of American Overseas Research Centres all provided vital resources or important contacts at crucial moments. Such institutions gave me more than parasites, of course, but these creatures have been a fitting way to explore a range of issues that go beyond cooperation and symbiosis.

Our own humble group at Environmental Humanities Switzerland has served as a reminder that people really do care about these issues no matter their official support or disciplinary background. My parasitic journey has also included crucial stops at archives assisted by dedicated archivists working at the u.s. National Parasite Collection; National Agricultural Library; Rockefeller Archive Center; Archivio di Stato di Cagliari; Istituto Superiore di Sanità; Istituto Etnografico della Sardegna; James C. Lovelock Papers, Wroughton; Frank Salisbury Papers, University of Utah Marriott Library Special Collections; and the who Archives, Geneva, among others.

In no particular order, I mention the following individuals who at various moments or opportunities gave of themselves, their ideas or their enthusiasm – even though few of them may have realized that I had parasites in mind while conversing with them: James McCann, Kate Wright, Christof Mauch, Anna Mazanik, Jim Webb, Davide Orsini, Christoph Kueffer, Marco Armiero, Joanne Bauer, Joel Rosenthal, Sevgi Mutlu Sirakova, Tomás Bartoletti, John McNeill, Eric Hinderaker, Bill Cronon, Luisa Passerini, Steve Tatum, Giuliano Pancaldi, Pierluigi Cocco,

Mario Macis, Drew Keeling, David Lowenthal, Flurin Condrau, Peter Truöl, Dan Tamir, Dan McCool, Peter Coates, Ken Vernick, Mario Coluzzi, Carlo Contini, Eva Veronesi, Samer Angelone-Alasaad, Eric Hoberg – and not to forget the dedicated bookmakers at Reaktion including Emma Devlin, Alex Ciobanu, Martha Jay, Helen McCusker, Fran Roberts, Sharon Laverick and especially Vivian Constantinopoulos; since bookmakers are also those who take down bets, they certainly did with this one. Lastly, I couldn't have done it without Aldo, Alex and Elena – the fearsome trio that put up with my disjointed ramblings while (usually) keeping me from taking myself too seriously. To all, I give my heartfelt thanks.

PHOTO
ACKNOWLEDGEMENTS

The author and publishers wish to express their thanks to the sources listed below
for illustrative material and/or permission to reproduce it:

AdobeStock: pp. 38 (Alessio Orrù), 39 (murasal), 192 (Ivan Canavera), 214 (AI
Generated by HadK), 238 (Jason); © 2011 Bizarro Studios, distributed by King
Features Syndicate, Inc.: p. 12; from Peter J. Brown, 'Cultural Adaptations to Endemic
Malaria and the Socioeconomic Effects of Malaria Eradication in Sardinia' (PhD
Dissertation, State University of New York at Stony Brook, 1979), reproduced with
permission: p. 32; CDC/Alexander J. da Silva, PhD/Melanie Moser: p. 43; photo
Betsey Dexter Dyer: p. 207; from L. W. Hackett, *Malaria in Europe: An Ecological
Study* (London, 1937): p. 142; photos Marcus Hall: pp. 71, 80; from Maurice C. Hall,
Control of Animal Parasites (Evanston, IL, 1936): p. 62; photo DJ Helmes, www.
findagrave.com: p. 181; JacobMrox/Resident Evil Wiki, CC BY-SA 3.0: p. 227; Library
of Congress, Prints and Photographs Division, Washington, DC: p. 96; Ettore Loi/
AFP/Getty Images: p. 21; The Metropolitan Museum of Art, New York: p. 249; from
Alberto Missiroli, 'La Malaria nel 1944 e misure profilattiche previste per il 1945',
Rendiconti dell'Istituto Superiore di Sanità, VII (1944): p. 28; National Agricultural
Library, Special Collections, Beltsville, MD: p. 49; National Library of Medicine,
Bethesda, MD: pp. 76 (Fred Soper Papers), 115, 178; National Museum of Health and
Medicine, Otis Historical Archives, Silver Spring, MD: p. 137; from William Ramesey,
*Helminthologia; or, Some Physical Considerations of the Matter, Origination, and
Several Species of Wormes* . . . (London, 1668): p. 54; from *Rapporto Mensile* ERLAAS
(October 1948): p. 149; from Francesco Redi, *Osservazioni . . . intorno agli animali
viventi che si trovano negli animali viventi* (Florence, 1684), photos Wellcome Library,
London: pp. 55, 56; from François Roudaire, *La mer intérieure africaine* (Paris, 1883):
pp. 218–19; Smithsonian Libraries, Washington, DC: p. 175; photos Wolfgang
Suschitzky, courtesy IsreSardegna: pp. 26, 40, 41, 193; © The Trustees of the British
Museum: p. 17; Università degli studi di Cagliari (Pierluigi Cocco Papers): p. 138;
University of Utah Libraries, Special Collections, Salt Lake City (Frank B. Salisbury
Papers): p. 241; U.S. Department of Energy: p. 217; The Walters Art Museum,
Baltimore, MD: p. 215; from N. B. Ward, *On the Growth of Plants in Closely Glazed Cases*
(London, 1852): p. 236; Wellcome Collection, London: pp. 87, 144 (Shell, Nucleus
Film Unit, 1949), 171 (Leandro Lemgruber, University of Glasgow, CC BY 4.0); from
George Hieronymus Welsch, *Exercitatio de vena medinensi, ad mentem Ebnsinae*

INDEX

Page numbers in *italics* refer to illustrations

asymptomatic parasitaemia 15, 31–3,
98, 109, 166, 183, 259
Atabrine (quinacrine) 136–7, 137,
138–9, 138, 146, 148
atmospheric carbon dioxide 217, 222,
233–4
Australia 74, 187, 196, 204

baby milk 259
Bacillus anthracis (anthrax
bacteria) 164
Bacillus thuringiensis israelensis
(Bti) 149
bacteria 9, 43, 52, 70–71
bacteriophages 183, 188–9
Beaver, Paul C. 52–3, 78, 268
bed nets 92
bedbugs 268
benefits
of curative parasites 103–6
of mind-altering parasites 106–12
of mosquitoes 177, 180–82
mutual aid between species 69
of mutualistic parasites 97–103
of parasites 10, 14–18, 43–4, 62–4,
70–71, 79–83, 113–20, 165–6,
198–9
of pathogenic parasites 86–97,
118–20
of Plasmodium 151–7, 170–75
bioprecipitation 201–2, 205–6, 209
biorepositories 253–4
biospherics and Biophere 2 209–10,
235–42, 238, 239, 241
birch mushroom (Fomitopsis
betulina) 81
Black Death see plague
Black Zones (of malaria) 25, 28, 29–31
Blanchard, Raphaël 67
blood cells 35–6, 42, 43, 94–5, 149–50
blood pressure 152–6
bloodletting 113, 129, 134
Blue Zones (of longevity) 22–4,
29–31
'botany of desire' 111
botflies 154, 251–2
brain 107, 109, 153
British 80, 90–91, 174

Bti (Bacillus thuringiensis
israelensis) 149
bubonic plague see plague
Buettner, Dan 22–4
Buffon, George-Louis Leclerc, comte
de 231–2, 233
bulbous growths 68, 210

canals see drainage
cancers 146, 225–6
cane toads 187
carbon capture 216–17, 217
carbon dioxide 217, 217, 222, 233–4,
244
carbon tetrachloride 59–60, 73
Caribbean 83, 85, 89–90, 94, 95–7, 179
Carson, Rachel 15, 230
Silent Spring 143–4
casu marzu (formaggio marcio) (rotten
cheese) 190–92, 192, 193, 255
cats 9, 15, 49, 107–8, 109, 117, 169
Cervus elaphus (red deer) 251–2
Ceylon 92
cheese 190–94, 192, 193, 266
cheese flies 190–94, 192
Chenopodium ambrosiodes (American
wormseed) 59
Chernin, Eli 52
chewing 250, 257, 260
Chikungunya 182, 186
children 80, 92, 93, 114
China 113, 114, 118, 125–8, 144, 216, 257
chloroplasts 190, 224
chloroquine 121–6
cinchona bark 15, 125, 133
cinchonism 126
climate change 214, 222, 231, 244
Clostridium difficile 257–8
co-culture 111–12
co-existence 13, 34–5, 64–5, 155, 157–8
co-extinction 45
collaboration 46, 65, 240
colonization
and creature exchange 83–9
and disease spread 176, 179
and immunity 160, 165
and macroparasites 162
and malaria remedies 125

mutualism
  and colonialism 83
  and Gaian theory 230–35
  and humans 229
  leading to parasitism 72–4, 97–103, 188–9
  as species interaction 43–4, 62–3, 65, 66–72, 77
myxoma virus 74, 196

natural selection 65–7, 74, 91, 164, 188, 266
Necator americanus (North American hookworm) 59–61, 259, 262
  see also hookworms
New York, United States 268
New Zealand 86
niche theory 217–18, 220
Nigeria 33, 90–91, 133, 174, 271
noosphere 214–15, 240
Notoedres cati 49
  see also mites
nuclear warfare 212, 216

obligate parasites 45, 169, 246
ocean resources decline 225
odours 171–2, 220
Okinawa, Japan 22, 30–31
Oostvaardersplassen landscape reserve (Netherlands) 264
organelles 70, 72–3, 183–5, 188–9
Ötzi the Iceman 81–2

Panama Canal 91–2, 177, 216
Parascaris equorum 49
  see also roundworms
parasitaemia, asymptomatic 15, 31–3, 98, 109, 166, 183, 259
parasite collections 48–50, 57, 58, 59
parasite cycles 42, 43
parasites, definition 13–14, 52–4, 56–8, 166–8
parasites within parasites 182–90
parasitic manipulation 117
parasitism
  and cooperation 69–70
  fear of 266
  in human affairs 45–6

and human health 13–16
leading to mutualism 72–4, 188–9
parasites as hosts, hosts as parasites 165–8
as species interaction 43, 44, 62, 65, 68
parasitology
  definition 52–3
  established scientific field 56–8
  and mission to kill parasites 58
  rise of the science of 64
  science of 9–10, 43–4, 46, 50–51
  term 67
parasitophobia 267–74
Paris, France 113, 268
Paris Green insecticide 25, 144–5, 146
pathogenic parasitism 77–8, 86–97, 118–20, 197–9, 271–3
pathogens 42–3, 44–5, 84–9, 228–9
pathos 42, 228–9
Pauly, Daniel 225, 226–7
permafrost, melting 244–5
perspective of parasites 161–3
pesticides 139–47
  see also DDT
petroleum 141, 145, 146
pets, domestic 107–8
pharmaceuticals 122, 134–9, 148–50, 150, 151
pilot fish 44
pinworms 62, 79–81, 80
  Enterobius vermicularis 8, 80
Piophila casei (cheese fly) 190–92
plague 104, 118–19, 163, 169
planetary symbiosis 202, 208, 221–5
plants 63, 68–9, 111, 128–9, 201–3, 232–3, 254
plasmids 183–4, 188–9
Plasmodium
  acquired immunity 91–2, 94–7
  and antimalarials 15, 132–3
  benefits of 95–6, 112, 119, 151–7, 170–75
  and blood pressure 152–6
  conservation 253
  and coronavirus 156
  as curative parasites 105